FRONTIERS IN ROBOTICS RESEARCH

Frontiers in Robotics Research

Max A. Denket
Editor

Nova Science Publishers, Inc.
New York

For permission to use material from this book please contact us:
Telephone 631-231-7269; Fax 631-231-8175
Web Site: http://www.novapublishers.com

LIBRARY OF CONGRESS CATALOGING-IN-PUBLICATION DATA
Available upon request.

1-60021-097-X

Published by Nova Science Publishers, Inc. ✢ New York

CONTENTS

PREFACE

Robotics began as a science fiction creation which has become quite real, first in assembly line operations such as automobile manufacturing, airplane construction etc. They have now reached such areas as the internet, ever-multiplying-medical uses and sophisticated military applications. Control of today's robots is often remote which requires even more advanced computer vision capabilities as well as sensors and interface techniques. Learning has become crucial for modern robotic systems as well. This new book brings together leading research in this exciting field.

As explained in Chapter I, industries must improve productivity and quality management continuously in order to keep an advantage over their competitors. In western countries the stress is put on the massive use of technology as labor costs are higher than in other countries. One of the most recent technologies is Machine Vision, used in industry for manufacturing, assembly and quality control. The visual inspection process might be defined as the one used to determine if a given product fulfills specifications, which include the absence of defects, such as lack (or excess) of material, homogeneous visual aspect, required color, predetermined texture, etc. This process requires the use of images that contain enough information about the visual appearance of the product to be inspected. It also requires a quick, reliable decision making step and a feedback loop to the manufacturing process. The use of vision systems in industry brings additional reliability where humans tend to make mistakes because of monotonous working processes. Most of the research done in the field of Machine Vision shows the efficacy of the identification algorithms without explaining how they can be integrated in the manufacturing process. In this chapter, visual inspection and quality control theory are combined to develop a robust intelligent inspection system within manufacturing applications.

The intelligent visual inspection system is applied to the control of continuous line production of metallic surfaces. The system is divided into the identification of each metallic surface and the control of the production line. The characterization of the visual aspect of individual metallic surfaces is studied using on-line quality control limits to quickly detect the presence or absence of defects in the image. This step is a hybrid between the two usual methods of pattern comparison and theoretical decision. It incorporates quality control theory to statistically model the pattern for defect-free products. Specifically, individual 6-sigma control limits are set so the inspection error is minimized. Whenever defects are identified, a classification is performed using texture analysis only over the areas with defects. It is based on the study of the frequencies obtained by the Fourier transform of these smaller areas.

Finally, the quality control of the whole production line is performed. Control charts for nonconformities are used to quantify the defects along time. Experimental results demonstrate the efficacy and robustness of the proposed methodology.

When conditions permit, experienced divers are invariably deployed to inspect underwater installations for damages to structure and weld joints, as presented in Chapter 2. With advances in Underwater Robotic Vehicles, unmanned inspection techniques promise to be a more cost effective and safe option. One of the major challenges to underwater robotic inspection is in the design of dexterous manipulators to operate in the underwater environment. The work environment can be highly constrained due to limited accessibility of the vehicle to the interconnected structures and the limited dexterity of existing systems. The design of special purpose, custom-made manipulators for such specialized applications is a challenging task. We present and discuss the design of a special purpose, custom made manipulator for inspection of underwater structures. A review of the state-of-art underwater manipulators is included. The functional requirements of an underwater manipulator for subsea inspection are discussed, and the desired performance requirements are identified.

The design challenges and developmental issues of a special purpose manipulator for underwater inspection are elaborated. A novel kinematic structure is proposed for Underwater Robotic Vehicle (URV) operation, having a 2 d.o.f launching stage and a 5 d.o.f inspection stage. In order to reap the benefits of both the hydraulic and electric actuators, a hybrid actuator-power structure has been adopted for the manipulator. Large torque and relatively lower positioning requirements of the launching stage dictate the use of high torque hydraulic actuators, for this stage. Similarly, relatively lower torque and high accuracy requirements of the forward inspection stage make the specifying of electrical actuators the obvious choice.

In underwater applications, reliability is a paramount consideration. Proportional electro-hydraulic servo valves are susceptible to particular contaminants in the hydraulic lines. The simpler poppet valve designs are more tolerant to such contaminants, but are used in a simple full-on or full-off ('bang-bang') arrangement. This lacks the opportunity for fine motion control. In our design, the actuating current to the poppet valve solenoid is 'pulse width modulated' to allow a more precise control of the hydraulic actuator. This is achieved through more sophisticated switching of control signals by the embedded micro controller. Another major source of failure for underwater systems is the connectors. The novelty of design extends to the inspection section by the integration of intelligent actuator modules. This provides for a simpler mechanical and electrical interconnection configuration, resulting in a reduction in cable lengths and number of connectors. Underwater connectors remain a major cause of equipment malfunction.

Design philosophy, system architecture and design constraints are presented and discussed. Results of simulation studies, Experimental analysis, and optimisation problems are also presented. The article serves as a technical research paper and provides insights into a detailed case study of the processes and considerations of realizing a novel special-purpose manipulator for underwater robotic vehicle (URV) applications.

Chapter 3 presents the band-pass quadrature filters which are extensively used in computer vision to estimate information from images such as: phase, energy, frequency and orientation, possibly at different scales and utilise this in further processing-tasks. Arguably, local phase is the most desirable information since it encodes most of the relevant perceptual information in an image. Phase-based image analysis techniques have been used for estimating optical flow, matching, registration and feature detection. The estimation is intrinsically noisy and depends critically on the choice of the quadrature filters. In this chapter, the authors first study the mathematical properties of the quadrature filter pairs most commonly seen in the literature and then consider some new pairs derived from the classical feature detection literature. In the case of feature detection, they present the first attempt to design a quadrature pair based on filters derived for optimal edge/line detection. A comparison of the filters is presented in terms of feature detection performance, wherever possible, in the sense of Canny and in terms of phase stability. The authors conclude with remarks on how their analysis can aid in the choice of a filter pair for a given image processing task.

In Chapter 4, the human interaction and the robot's autonomy which are key functions that can spread the use of the robot in human daily life is studied. The remote interaction is a special type of human-robot interaction, where the human and the robot are separated by physical barriers but linked via telematics technologies (= telecommunication + informatics). Due to the nature of intelligence needed for the modern robots, remote interaction systems must combine the dream of the artificial intelligence field of creating an intelligent being, creature, or program with human-computer interaction's vision of building systems that are centered around the user's needs and desires. This introduces a number of technical challenges, which must be addressed such as local autonomy, control strategy, communication medium constraints, feedback methods and control interfaces. In this chapter, these considerations are discussed. The chapter also presents a multilayer architecture, which can be used to facilitate the remote interaction with mobile robots using different interaction elements for research and educational purposes.

Chapter 5 will conclude that the increasing demand for robots in various fields of application has created a thirst, to make them intelligent, interactive, cooperative and emotional as well. To reach this milestone, a lot of research is yet to be carried out, in different fields of robotics research. Motion planning and coordination of multiple robots may be considered as one of such important fields. This chapter concentrates on this field of robotics research, discusses different possible schemes using both conventional algorithmic as well as emerging soft computing approaches. It has been pointed out that the fuzzy logic and/or neural network-based controller evolved using an artificial evolution tool like genetic algorithm, could be the candidate adaptive robot controller.

In: Frontiers in Robotics Research
Editor: Max A. Denket, pp. 1-33
ISBN: 1-60021-097-X

Chapter 1

INTELLIGENT VISUAL INSPECTION SYSTEM FOR STATISTICAL QUALITY CONTROL OF A PRODUCTION LINE

José Mª Armingol*, Javier Otamendi, Arturo de la Escalera, José Manuel Pastor and Francisco José Rodríguez

Intelligent Systems Lab
Department of Systems Engineering and Automation
Universidad Carlos III de Madrid
C. Butarque 15, 28911 Leganés, Madrid, Spain

Abstract

Industries must improve productivity and quality management continuously in order to keep an advantage over their competitors. In western countries the stress is put on the massive use of technology as labor costs are higher than in other countries. One of the most recent technologies is Machine Vision, used in industry for manufacturing, assembly and quality control. The visual inspection process might be defined as the one used to determine if a given product fulfills specifications, which include the absence of defects, such as lack (or excess) of material, homogeneous visual aspect, required color, predetermined texture, etc. This process requires the use of images that contain enough information about the visual appearance of the product to be inspected. It also requires a quick, reliable decision making step and a feedback loop to the manufacturing process. The use of vision systems in industry brings additional reliability where humans tend to make mistakes because of monotonous working processes. Most of the research done in the field of Machine Vision shows the efficacy of the identification algorithms without explaining how they can be integrated in the manufacturing process. In this chapter, visual inspection and quality control theory are combined to develop a robust intelligent inspection system within manufacturing applications.

The intelligent visual inspection system is applied to the control of continuous line production of metallic surfaces. The system is divided into the identification of each metallic surface and the control of the production line. The characterization of the visual aspect of individual metallic surfaces is studied using on-line quality control limits to quickly detect the presence

* E-mail address: armingol@ing.uc3m.es

or absence of defects in the image. This step is a hybrid between the two usual methods of pattern comparison and theoretical decision. It incorporates quality control theory to statistically model the pattern for defect-free products. Specifically, individual 6-sigma control limits are set so the inspection error is minimized. Whenever defects are identified, a classification is performed using texture analysis only over the areas with defects. It is based on the study of the frequencies obtained by the Fourier transform of these smaller areas. Finally, the quality control of the whole production line is performed. Control charts for nonconformities are used to quantify the defects along time. Experimental results demonstrate the efficacy and robustness of the proposed methodology.

Key words: Quality Control Charts, Automated visual inspection, Machine Vision, Statistical Pattern Recognition, Steel Surfaces

1 Introduction

Automated visual inspection tries to use both Computer Vision and Artificial Intelligence disciplines to control the visual quality of industrial products. In repetitive and monotone pictures, the inspection tasks carried by human operators are tiring and tedious. Besides, in today's open market environment, visual quality control is, in many cases, decisive to be competitive. For these main reasons, in the last couple of decades, the number of scientific publications related to Automated Visual Inspection Systems has grown significantly to show how human visual inspection is being replaced by computer vision systems.

Visual Inspection Systems improve the quality control of inspection tasks within the production process. Some of the main applications of visual inspection systems are: precision measurements, structural defects, pattern modeling. Some of the benefits that visual quality control incorporates in the inspection process are:

Speed: quicker and more reliable than humans. It performs high speed checks without stopping and affecting the flow of the production line.

Repeatability: high precision, consistent systems

High resolution cameras: which detect structural defects that the human eye is not capable of seeing due to its inferior resolution characteristics.

Security: inspections might be carried out eliminating operators risk.

Cost reduction: it reduces the production costs due to the reduction in spoilage and reprocessing via a quicker detection of faults in the processes.

The inspection process is defined as [14] the one that is followed to decide whether a given product meets previously established specifications, which are defined as the quality standard. In the case of visual inspection, these specifications include the absence of defects such as lack of material, homogeneous visual aspect, required color, predetermined texture, etc. This process requires the use of images which include enough information about the visual appearance of the product to be inspected. It also requires a quick, reliable decision making step and a feedback loop to the manufacturing process. Besides traditional inspection in mass production with statistical methods, and with the aim of eliminating producer's and buyer's risks, sometimes it is necessary to perform a 100% check of production [27]: for example, in industries like ceramics, steel, etc., due to competitive factors, or in industries like pharmaceuticals, aerospace, etc., due to critical factors [7].

The automation of visual inspection implies including in the systems a high degree of reasoning capacity in the decision making process or in the classification of the defects regarding its visual appearance, eliminating then the lack of homogeneity in the inspection results due to human factors, therefore increasing efficacy [24] [8].

Visual inspection techniques are classified in two groups. The first one includes comparisonwith-patterns techniques, that is, the defects are detected by comparing defect-free patterns with images of actual products. The second group relates to theoretic decisions, used to verify a set of characteristics of the product against those of an ideal product.

- Comparison with a Pattern

This technique is based on the comparison pixel by pixel of two images, one of the inspected product and another of the ideal product. This last one must include the visual model of the product to be studied. In the case of multilevel images, patterns must be carefully developed to assure that small defects are detected and not confounded with the inherent variability of the visual characteristics of the product [25] and the errors in the measuring process [23]. This type of technique is sufficiently flexible, and sometimes is the only possible way to perform a reliable inspection. It has been used in the quality control of printed circuit boards [18], tiles [22] [6], etc.

- Theoretic Inspection

This type of methods is based on the adequate parameterization of the defect-free product. This parameterization consists of a vector of image characteristics, like areas, perimeters, shapes, moments, textures, histograms, etc. It is necessary to develop a classifier which helps decide on the quality of the inspected product (based on rules and/or neural networks) [13]. This type of technique has been used in the quality control of metallic pieces [27], paper pulp [11] [17], etc.

Inspection processes based on textures are included in this group [21]. These are processes in which a visual pattern with a certain degree of uniformity is analyzed [2]. The texture allows for the representation of the properties of the product and the localization of the defects.

In this chapter, a methodology is presented for the identification of defects in metallic surfaces. The proposed method is a hybrid between the inspection with a pattern and the use of theoretic decisions, with the aim to minimize the influence of their disadvantages and to maximize their advantages. The objective of this inspection process is to identify possible defects, specifically bumps and scratches, in metallic surfaces with a uniform degree of curvature, as depicted in figure 1. The detection and quantification of defects is to be done online in the manufacturing process.

Figure 1. Inspection process.

The system is divided into the identification and classification of each defect on the metallic surfaces (Section 2) and the control of the production line (Section 3). The characterization of the visual aspect of individual metallic surfaces is studied using on-line quality control limits to quickly detect the presence or absence of defects in the image. This step is a hybrid between the two usual methods of pattern comparison and theoretical decision. It incorporates quality control theory to statistically model the pattern for defect-free products. Specifically, individual control limits with 6-sigma limits are set so the inspection error is minimized. Whenever defects are identified, a classification is performed using texture analysis only over the areas with defects. It is based on the study of the frequencies obtained by the Fourier transform of these smaller areas. Finally, the quality control of the whole production line is performed. Control charts for nonconformities are used to quantify the defects along time. Experimental results demonstrate the efficacy and robustness of the proposed methodology.

2 Visual Inspection System

Choosing the illumination system constitutes one of the fundamental steps in any visual inspection system, as it is used to pinpoint the undesirable characteristics that must be detected during inspection. Therefore, the illumination system is to possess a very good spatial and temporal uniformity.

The illumination system is crucial when dealing with surfaces which are very reflective and have a plain texture, as in the case of metallic surfaces. The defects related to reflection are of superficial nature whereas the ones related to texture are due to design requirements.

For the detection of both defects, it is essential to optimally locate the sources of light and the video camera. Only if the object is correctly illuminated, errors like changes in tone or contrast might be avoided.

The light that gets to the object should provide constant intensity levels over the whole image. Whenever it is necessary to point out microstructure alterations of the surface, the illumination system is to guarantee, besides uniformity, the correct orientation of the incident light. In these cases, the camera must capture the specular component of the reflected light.

This component is different from the diffused light in terms of the intensity levels and the orientation of incident light.

In large objects, where directional illumination of great intensity is used, the area that is close to the source of light shows a higher level of gray than those that are further away, as shown in Figure 2.

Figure 2. Directional illumination in metallic surfaces.

In Figure 2, the light distribution is not uniform. In Section 2.1, an algorithm that models the characterization of the visual aspect using statistical techniques is proposed. The texture of the surface has been modeled with Gaussian distributions using sequences of defect-free images. The algorithm allows for alterations in the microstructure of the surface in terms of the changes in the angles of reflection of the rays of light coming from the source of illumination, with relation to the predefined pattern. The model includes the effect of the noise associated to the transmission of the signal and to the source of light used.

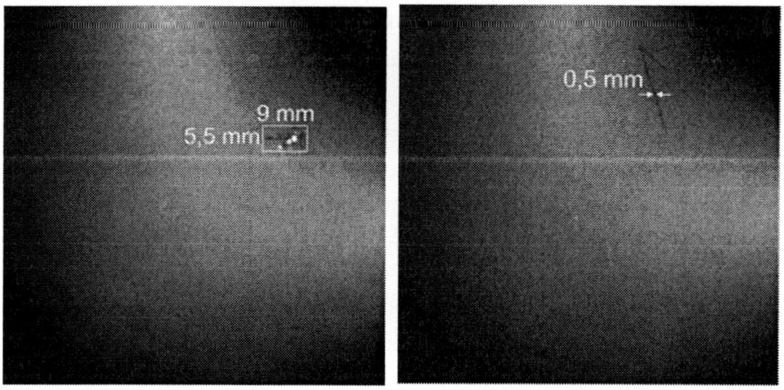

(a) (b)

Figure 3. Defects in the metallic surface.

Figure 3 depicts two different defects. In Figure 3(a), the plate has a small bump which induces deformations in its structure within 9x5.5 mm, whereas in Figure 3(b), a scratch of approximately 0.5 mm is shown.

In Section 2.2, an algorithm for the quantification and classification of this kind of defects is described. It is based on the application of the Fourier transform to analyze the texture of the surfaces.

2.1 Detection of Defects

The detection algorithm is used to determine if the product meets specifications or defects are found on its surface. It is based on the development of useful patterns that represent defect-free products, patterns which are used to statistically determine the quality of the product.

The original idea of this chapter is to combine visual inspection with quality control techniques that are commonly used in on-line inspection of manufacturing systems. Usually, quality control limits are used by operators to monitor performance characteristics that are measured by traditional methods (gages, weight scale, etc.). In vision systems, the characteristics are incorporated into the charts by automatically converting an image into a measurable value (gray scale). On-line inspection is then more flexible and robust.

A posterior image processing step has also been developed to correctly determine the location of the defect, if it has been pinpointed by the control limits.

2.1.1 Statistical Pattern Modeling

The model is based on quality control theory, more specifically, on statistical confidence intervals. The philosophy behind the intervals is equivalent to the philosophy used to process images for inspection purposes. Pixels will be acceptable as long as their gray level stays within certain acceptable limits. Statistics allows for a robust design of these quality control (QC) limits so inspection errors are minimized.

2.1.1.1 Development of Quality Control Limits

QC limits are used to monitor the quality of a given process over time, both in terms of its mean behavior and its variability. They are simple analytical tools whose objective is to determine if the process is in statistical control, that is, if the mean value of the variable and its variation continue to lie within certain boundaries, called control limits.

To create the confidence interval, an initial sample of at least 20 values is necessary (preferably more than 30). The mean $(x\ )$ and standard deviation (s) of the sample are calculated. The limits are set as the mean value plus or minus k standard deviations (usually k=3):

$$(x - ks, x+ks) \tag{1}$$

Two further comments on how to apply the confidence limits are necessary at this point [16]. First, the sample data must be checked for normality. Second, there exists the possibility of committing inspection errors. The following two sections analyze in detail these comments.

2.1.1.2 Test for Normality

One of the assumptions included in quality control theory is that the underlying random variable is normal, so a prior check for normality is required.

The test for normality is used to discriminate between the following two hypotheses:

- H_0: data follow a normal distribution.
- H_1: data do not follow a normal distribution.

The result of performing the test is a p-value which determines the significance level at which H_0 is rejected. If the required significance level is less than the calculated p-value, H_0 is not rejected, whereas if the level is greater than the p-value, H_0 is rejected. Usual significance levels for a single test are 0.05 and 0.01 (95% and 99% confidence).

The Anderson-Darling test is one of the more powerful tests for normality found in the literature [5]. The test is valid for small samples and calculates a statistic called A_2, which is calculated as follows:

Arrange the sample in ascending order

$$x_{(1)} \leq \ldots \leq x_{(i)} \leq \ldots \leq x_{(n)}$$

Calculate standardized values $y_{(i)}$

$$y_{(i)} = \frac{x_{(i)} - \bar{x}}{s}$$ where x is the mean and s is the standard deviation of the sample

Calculate P_i from the tables of the cumulative distribution function of the standard normal

$$P_i = \Phi(y_{(i)})$$

Compute A_2

$$A^2 = -\frac{[\sum_{i=1}^{n}(2i-1)(In\mathrm{P}_i + In(1-\mathrm{P}_{n+1-i}))}{n} - n$$

Compute the modified statistic A*

$$A^* = A^2(1\ .0+0.75/n+2.25/n^2)$$

Calculate the p-value

If $A^*<0.2$

ln p-value = -13.436 + 101.14 A^* - 223.73 $(A^*)^2$

Elseif $A^*<0.34$

ln p-value = -8.3 18 + 42.796 A^* - 59.938 $(A^*)^2$

Elseif $A^*<0.6$

ln p-value = 0.9177-4.279 A^* -1.38 $(A^*)^2$

Else

    ln p-value = 1.2937-5.709 A* + 0.0 186 $(A^*)^2$

Endif

If A^*<0.34

    p-value = 1-EXP(ln p-value)

Else

    p-value = EXP(ln p-value)

Endif

Reject Ho if the required significance level is greater than the p-value

For the visual system, and for same pixel in twenty images, the test for normality is shown in Table 1. Ho is not rejected for significance levels that are less than 0.2921, so the underlying distribution of the pixel follows a normal distribution.

2.1.1 3 Sampling Errors

By using quality control techniques charts there exists the possibility of committing decision errors, both of rejecting good products and accepting bad ones. Let's quantify these important possible errors, which are directly related to the chosen value of *k* (number of standard deviations used to calculate the limits).

Table 1. Test for normality.

												A2	A*	ln p-value
Mean	121,50					Mean	121,50					0,4215	0,4397	-1,2307
STD	5,53					STD	5,53					-408,43		
i	X	Y	P	1-P	ln P	n+1-i	X	Y	P	1-P	ln (1-P)			
1	110,00	-2,08	0,0188	0,9812	-3,97	20	131,00	1,72	0,9571	0,0429	-3,15	-7,12		
2	114,00	-1,36	0,0875	0,9125	-2,44	19	129,00	1,36	0,9125	0,0875	-2,44	-14,62		p-value
3	116,00	-0,99	0,1600	0,8400	-1,83	18	127,00	0,99	0,8400	0,1600	-1,83	-18,33		0,2921
4	116,00	-0,99	0,1600	0,8400	-1,83	17	126,00	0,81	0,7921	0,2079	-1,57	-23,82		
5	117,00	-0,81	0,2079	0,7921	-1,57	16	126,00	0,81	0,7921	0,2079	-1,57	-28,27		
6	117,00	-0,81	0,2079	0,7921	-1,57	15	125,00	0,63	0,7366	0,2634	-1,33	-31,95		
7	118,00	-0,63	0,2634	0,7366	-1,33	14	125,00	0,63	0,7366	0,2634	-1,33	-34,69		
8	119,00	-0,45	0,3256	0,6744	-1,12	13	125,00	0,63	0,7366	0,2634	-1,33	-36,84		
9	119,00	-0,45	0,3256	0,6744	-1,12	12	124,00	0,45	0,6744	0,3256	-1,12	-38,15		
10	122,00	0,09	0,5360	0,4640	-0,62	11	124,00	0,45	0,6744	0,3256	-1,12	-33,17		
11	124,00	0,45	0,6744	0,3256	-0,39	10	122,00	0,09	0,5360	0,4640	-0,77	-24,40		
12	124,00	0,45	0,6744	0,3256	-0,39	9	119,00	-0,45	0,3256	0,6744	-0,39	-18,12		
13	125,00	0,63	0,7366	0,2634	-0,31	8	119,00	-0,45	0,3256	0,6744	-0,39	-17,49		
14	125,00	0,63	0,7366	0,2634	-0,31	7	118,00	-0,63	0,2634	0,7366	-0,31	-16,51		
15	125,00	0,63	0,7366	0,2634	-0,31	6	117,00	-0,81	0,2079	0,7921	-0,23	-15,62		
16	126,00	0,81	0,7921	0,2079	-0,23	5	117,00	-0,81	0,2079	0,7921	-0,23	-14,45		
17	126,00	0,81	0,7921	0,2079	-0,23	4	116,00	-0,99	0,1600	0,8400	-0,17	-13,44		
18	127,00	0,99	0,8400	0,1600	-0,17	3	116,00	-0,99	0,1600	0,8400	-0,17	-12,20		
19	129,00	1,36	0,9125	0,0875	-0,09	2	114,00	-1,36	0,0875	0,9125	-0,09	-6,78		
20	131,00	1,72	0,9571	0,0429	-0,04	1	110,00	-2,08	0,0188	0,9812	-0,02	-2,45		

- *Errors per variable (per pixel)*

The primary error that results from the use of a confidence interval in each sample is the possibility of determining that the process has gone out of control when indeed the process remains in control. This error is usually called á. It is the probability that the new sample falls outside the limits if the process stays in control, so its value is:

$$\alpha=\Pr(z>|ks|)=2*\Pr(z>ks) \qquad (2)$$

where z is the standard normal distribution, which is included in tabular form in statistics textbooks. It is assumed that the variables included in the graph follow this distribution, as the normal causes or white noise also follow the normal or Gaussian distribution. For example, if k=3, the required error is á=0.27%.

The other possible error, called â, is the probability that the process stays stable when indeed is out of control. Its calculation is more complex, as it is necessary to know the new normal distribution that the process follows. The size of the errors varies in opposite directions. As k increases, á decreases while â increases.

- Errors per study (per image)

If M variables are being monitored in parallel, the total error is a function of the individual errors on each variable. If the confidence intervals are considered independent, the total error is calculated as follows:

$$\text{Error} = 100*[1-(1-\alpha)^M] = 100*[1-(1-2*\Pr(z>ks))^M] \quad (3)$$

The following graph (Figure 4) shows the relationship between the scalar k and the sampling error, for various levels of M, ranging from 5 12x5 12 to 1024x1024 pixels (current best precision) and beyond:

For the total error to be less than 10%, k must be at least 5. If k=6, the error vanishes for any M. In the case of the vision system, the á error is the possibility of detecting a fault when the object is correct, and the â error is the possibility of omitting a fault. The total number of charts is the number of pixels in which the image has been coded. The choice of k is 6.

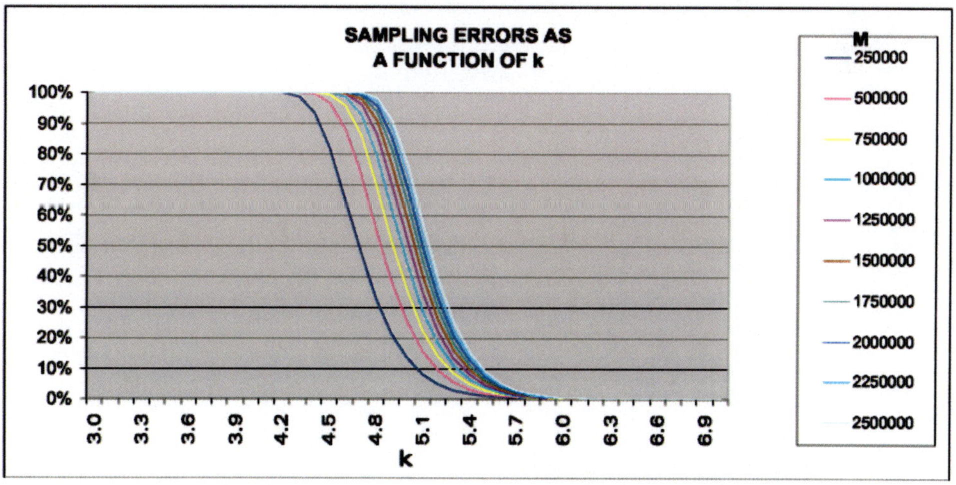

Figure 4. Relationship between k and the sampling error. 2.1.1.4 Pattern model.

The pattern is therefore an image made of M pixels, each with its confidence limits, with the values measured in the gray scale. To calculate the acceptance limits, a significant number of images (t 20) are taken of error-free metallic surfaces. Values for the mean and the standard deviation for each pixel are calculated and a normality check is performed. Then the limits are set at the mean value plus or minus k standard deviations. Since the total number of charts is M, to reduce the possibility of committing an á error the choice of k is 6.

2.1.2 Surface Inspection.

The algorithm must convert the image obtained with the video camera and compare it with the given pattern. Defects should be found with accuracy and located in the metallic surface.

2.1.2.1 Adjustment for Illumination

For a given surface, images are obtained with a video camera. The images are divided into pixels, each of which is assigned a value in the gray scale. These values, however, might be distorted depending on the positioning of the robot. A transformation of the gray values should be performed in the case of a flaw in the source of light or the positioning of the camera.

The algorithm incorporates a step to check for adequate illumination and performs a shift in the value if necessary. An adaptive process based on window analysis has been developed. For a given window, the difference between the value of the pattern and the processed image is calculated, for each pixel included in the window. The distribution of the differences is assumed to be normally distributed and estimated by estimating its mean value and standard deviation. If no abnormal individual difference is found, a shift is performed by moving the value of each pixel of the inspected image by a fixed amount, which is the difference in the mean value for each window. If, on the other hand, any of the differences is too large, the shift is not performed and the abnormal values are maintained for posterior checks.

The algorithm is the following:

1. Select a window with P pixels
2. For any pixel p:

   Calculate x_p: gray value for the pixel in the pattern
   Calculate y_p: gray value for the pixel in the inspected image
   Calculate the difference between the two readings:

$z_p = x_p - y_p$

Table 2. Adjustment for illumination.

			X					Mean	STD
162.0	158.8	158.9	156.0	153.6	153.4	152.6	151.8		
161.0	156.7	155.2	152.4	150.2	150.0	150.2	151.1		
157.7	155.0	151.5	149.1	147.5	146.1	147.5	149.3		
154.7	153.1	150.7	147.3	145.9	144.4	145.9	147.4		
153.5	151.5	151.2	147.0	145.5	142.8	144.2	144.1		
152.5	150.9	152.0	146.7	144.8	142.2	142.1	140.8		
150.7	149.1	150.5	143.9	142.0	140.1	139.1	137.1		
148.4	147.0	147.1	141.6	140.1	139.9	140.2	139.5	148.67	5.76
			Y						
144.0	143.0	143.0	143.0	143.0	143.0	143.0	146.0		
144.0	143.0	141.0	144.0	142.0	143.0	141.0	144.0		
145.0	141.0	143.0	141.0	142.0	141.0	143.0	142.0		
142.0	141.0	143.0	141.0	142.0	141.0	143.0	144.0		
142.0	143.0	141.0	142.0	141.0	141.0	141.0	143.0		
143.0	143.0	143.0	143.0	142.0	142.0	141.0	141.0		
143.0	141.0	145.0	141.0	143.0	141.0	141.0	141.0		
142.0	140.0	143.0	142.0	143.0	139.0	142.0	142.0	142.28	1.28

Z

18.0	15.8	15.9	13.0	10.6	10.4	9.6	5.8		
17.0	13.7	14.2	8.4	8.2	7.0	9.2	7.1		
12.7	14.0	8.5	8.1	5.5	5.1	4.5	7.3		
12.7	12.1	7.7	6.3	3.9	3.4	2.9	3.4		
11.5	8.5	10.2	5.0	4.5	1.8	3.2	1.1		
9.5	7.9	9.0	3.7	2.8	0.2	1.1	-0.2		
7.7	8.1	5.5	2.9	-1.0	-0.9	-1.9	-3.9		
6.4	7.0	4.1	-0.4	-2.9	0.9	-1.8	-2.5	6.39	5.26

	Confidence Limits	
t	-6	6
	-25.14	37.92

D

0.0	0.0	0.0	0.0	0.0	0.0	0.0	0.0	
0.0	0.0	0.0	0.0	0.0	0.0	0.0	0.0	
0.0	0.0	0.0	0.0	0.0	0.0	0.0	0.0	
0.0	0.0	0.0	0.0	0.0	0.0	0.0	0.0	
0.0	0.0	0.0	0.0	0.0	0.0	0.0	0.0	
0.0	0.0	0.0	0.0	0.0	0.0	0.0	0.0	
0.0	0.0	0.0	0.0	0.0	0.0	0.0	0.0	
0.0	0.0	0.0	0.0	0.0	0.0	0.0	0.0	SHIFT

Y'

50.39	149.39	149.39	149.39	149.39	149.39	149.39	152.39		
50.39	149.39	147.39	150.39	148.39	149.39	147.39	150.39		
51.39	147.39	149.39	147.39	148.39	147.39	149.39	148.39		
48.39	147.39	149.39	147.39	148.39	147.39	149.39	150.39		
48.39	149.39	147.39	148.39	147.39	147.39	147.39	149.39		
49.39	149.39	149.39	149.39	148.39	148.39	147.39	147.39		
49.39	147.39	151.39	147.39	149.39	147.39	147.39	147.39		
48.39	146.39	149.39	148.39	149.39	145.39	148.39	148.39	148.67	1.28

3. Calculate $\bar{z}$, s_z
4. Calculate the confidence interval on the difference

$$\left(\bar{z} - t_{p-1,\alpha/2} s_z, \bar{z} + t_{p-1,\alpha/2} s_z\right)$$

5. For any pixel p, calculate D:

   If z_p within the confidence limits

   $D_p=0$

   Else

   $D_p= 1$

6. Perform the adjustment:

   If every $D_p=0$

   Perform the shift and calculate y'_p

   $$y'_p=y_p+\bar{z}$$

   Else

   Do not perform the shift

Table 2 shows the application of the algorithm to the inspection system, for a 8x8 window. 2.1.2.2 Processing of the image

Figure 5(a) shows the image taken from the metallic surface after applying the adjustment for illumination. Figure 5(b) shows the result of comparing figure 5(a) pixel by pixel with the pattern and its corresponding control limits established with 6 standard deviations. In figure 5(b) black pixels are those whose gray level is within the control limits, whereas white pixels are those whose gray level is outside the control limits.

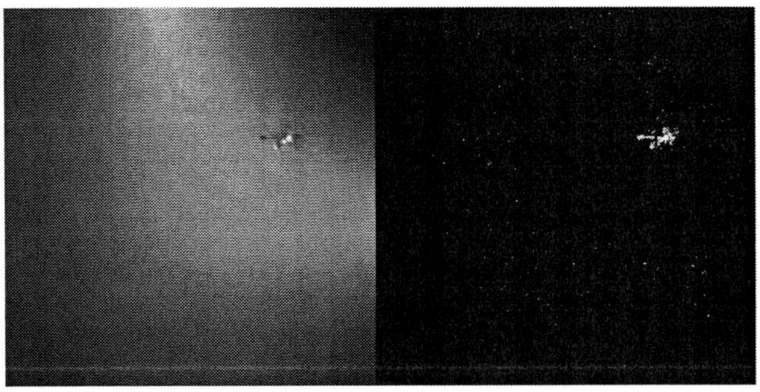

(a) (b)

Figure 5. Processing of the image.

The defect is clearly identified, but also isolated white pixels appear all over the processed image. The existence of the latter is due to dust particles that are included in the surface, which are almost unavoidable. The purpose of this processing step is to differentiate between the isolated pixels and the ones that correspond to defects.

To eliminate these isolated pixels, a morphologic filtering step is carried out over the whole domain [20], with the following kernel:

0	1	0
1	0	1
0	1	0

With this kernel, the number of neighbors is analyzed considering that the signaled pixels will show no more than two neighbors, whereas pixels with real defects will have at least three neighbors. The result of the processing step is shown in figure 6.

Figure 6. Application of the Kernel.

Figure 7 shows the histogram resulting from counting the pixels of figure 7. Five levels of gray are clearly shown, corresponding with the different number of neighbors.

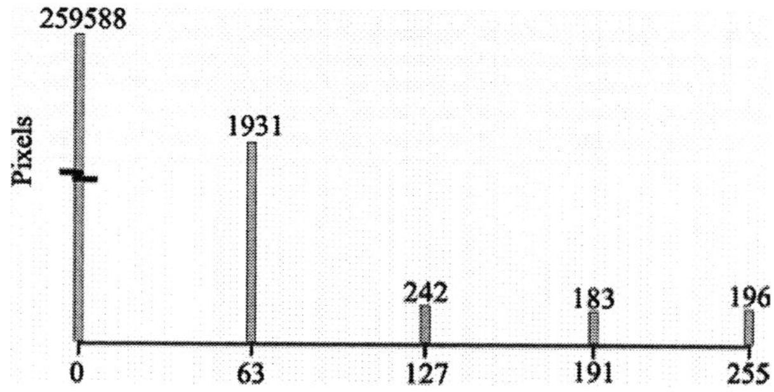

Figure 7. Levels of gray.

Figure 8 shows the image after applying the threshold of 3 or more neighbors to figure 7.

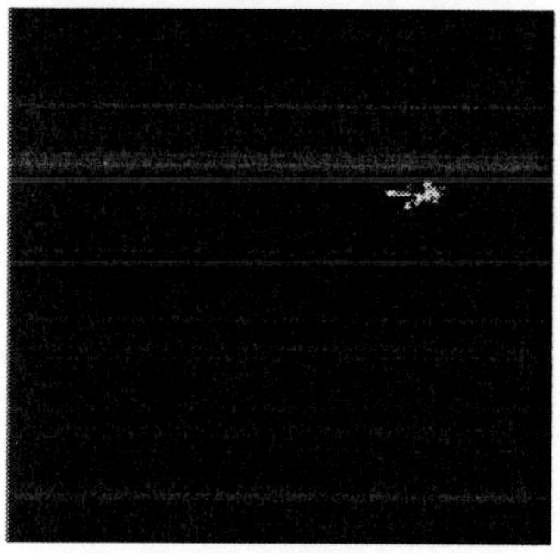

Figure 8. Image after threshold.

Finally, the white pixels are grouped and labeled to identify the defect. This defect is located with respect to the surface coordinates, also indicating its area, length and height. In the next section the found defects are further classified.

2.2 Quantification and Classification of Defects

So far, the visual inspection system described has two main advantages:

1. It quickly detects the presence of disconformities.
2. It is not affected by changes of illumination.

However, it also presents two clear limitations:

1. It does not perform a classification of the defects that are present in the image, just obtaining information to discriminate between defect-free surfaces and defective ones.
2. Since the metallic surfaces have different orientations respect to the camera, light reflections are fairly frequent. Those reflections are likely to saturate the sensor in large areas, achieving the maximum value in some pixels of the image and producing false alarms by classifying the surfaces as non-conforming when they really can be free of errors.

2.2.1 Texture Analysis

What follows is a discussion on how to manage the above limitations. The basic idea is the metallic surface is manufactured with a material that is uniform with respect to illumination.

This uniformity or texture might be modeled using the Fourier transform [9]. First, the transform of the image $f(x,y)$ is obtained:

$$F(u,v)=\frac{1}{N}\sum_{x=0}^{N-1}\sum_{y=0}^{N-1}f(x,y)e^{\frac{-j2\pi(ux+vy)}{N}} \tag{4}$$

to calculate the modulus $P(u,v)$:

$$P(u,v)=\left|F(u,v)\right|^2 \tag{5}$$

which expressed in polar coordinates, $P(r,è)$, helps derive the two functions that represent the texture:

$$D(r)=\sum_{\theta=0}^{\pi}P(r,\theta) \tag{6}$$

$$O(\theta)=\sum_{r=0}^{N/2}P(r,\theta) \tag{7}$$

$$r=\sqrt{u^2+v^2} \quad \theta=\arctan\frac{v}{u} \tag{8}$$

The values obtained using equation 6 represent the size of the dominant texture, whereas the one obtained using equation 7 add the direction (Figure 9). For the case included in this chapter, the analysis is based on the study of the variations in the values obtained with equation (7) when nonconformities appear, since the dominant texture obtained by equation (6) hardly ever changes even if there were many defects in the image because the predominance of the defect-free part.

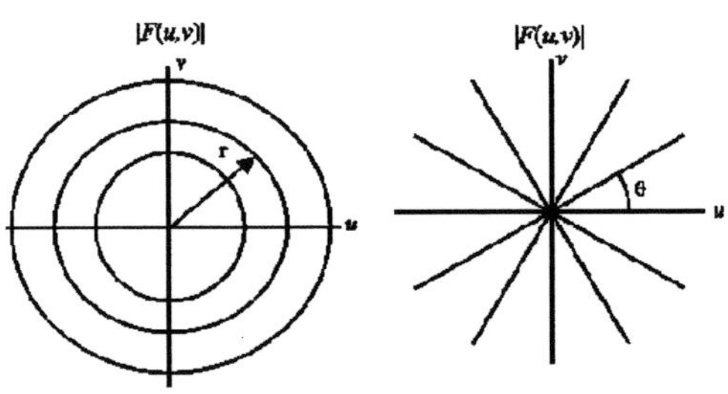

(a) (b)

Figure 9. Texture characterization in the frequency domain. (a) Graphical representation of function $D(r)$ used to obtain the size of the dominant texture (b) Graphical representation of function O(è) used to obtain the direction of the dominant texture.

Figure 10 presents four images of the same defect under different circumstances. If image 1 0a is taken as the base image, 10b presents a change of illumination, and as the defect can be seen from another point of view, 1 0c presents a pure translation and 1 0d a rotation. The second row shows the modulus of the corresponding discrete fast Fourier transform (DFFT) of each of the images. It must be highlighted that for the first cases the results are the same, whereas for the last one, the result is similar, although the DFFT of the image have been rotated in the same proportion. Figure 10e shows the dominant texture directions, following equation (7). It might be concluded then that the analysis in the frequency domain characterizes the nonconformity even with changes in location, rotation and illumination.

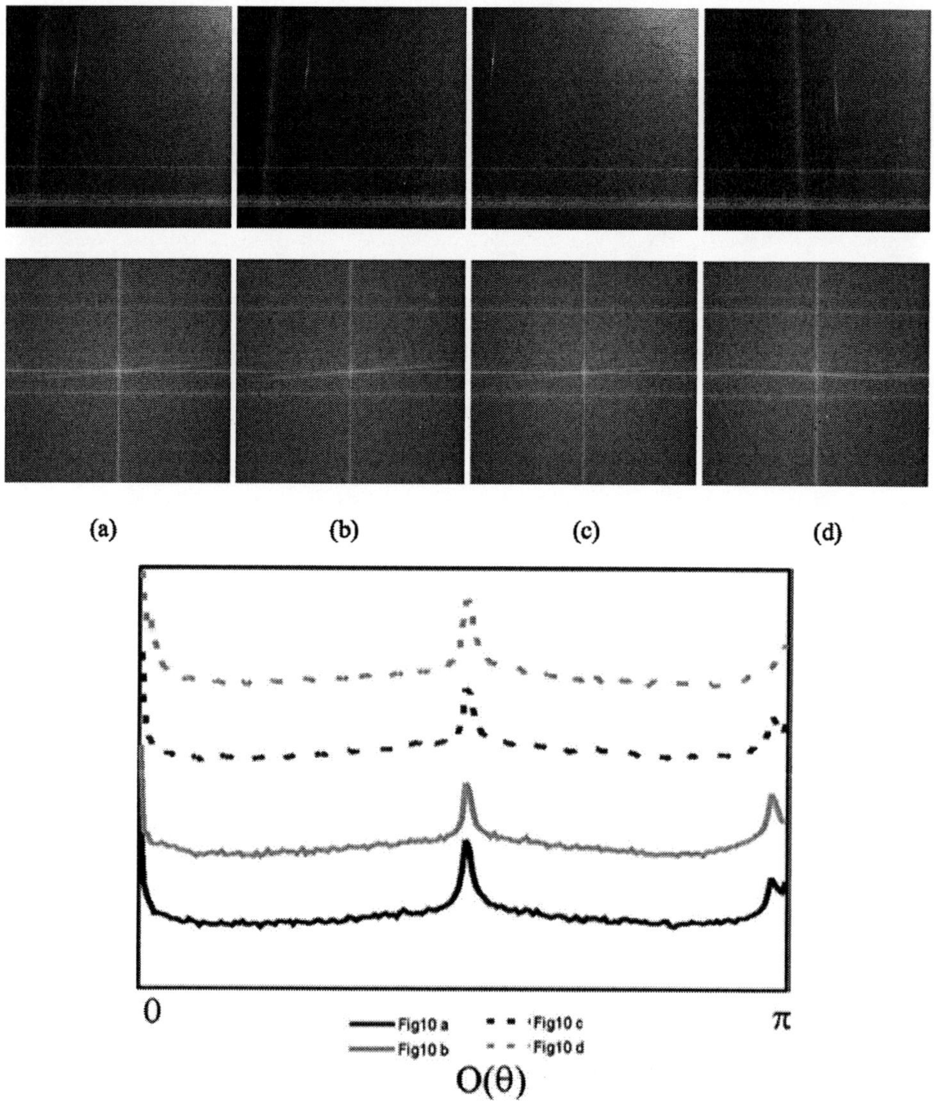

(e)

Figure 10. Properties of the discrete fast Fourier transform of some images. (a) The original image (b) Lighting Change (c) Translation (d) Rotation (e) Dominant texture directions.

2.2.2 Defects Location

The analysis based on the DFFT shows the presence of a defect in the image, but not its quantity or its location. The previous series of images have shown that the result is the same regardless of the position of the camera. Therefore, it is not possible to determine the location of the defect. Also, if the surface presents two different but identical defects, the appearance would be similar to having just one, due to the invariance against translation. Therefore, it is not possible to determine the number of defects. How might that information be obtained? The answer is to split the image into several equal-size sub-images. For example, let's take image 1 1a, which presents a defect-free surface. The DFFT is obtained (1 1b), and as the resolution of the texture has to be calculated, the initial image is divided into four equal-size sub-images, obtaining then the DFFT (1 1c) for each one of them. It can be seen that the textures described by equation (7) are similar for both resolutions. If the image is divided again and the DFFT (11d) performed, the results are the same. Therefore, it must be concluded that one image (or sub-image) is defect-free if O(è) has two maxima, one in 0 (or π) and the other in $\pi/2$ radians. What happens when there are nonconformities? Following the series of images with 11e, its DFFT is quite different since it has two additional local maxima, one per defect. If, like in the previous case, four sub-images are obtained, the results are different because three sub-images have a texture with only two maxima, and therefore are defect-free and there is one sub-image with four local maxima. This helps discriminate between zones of the image with defects and zones with a perfect appearance. Hence, the texture of an image of uniform appearance does not show any predominant direction, but as soon as one defect is present it is shown in the texture through a predominant direction.

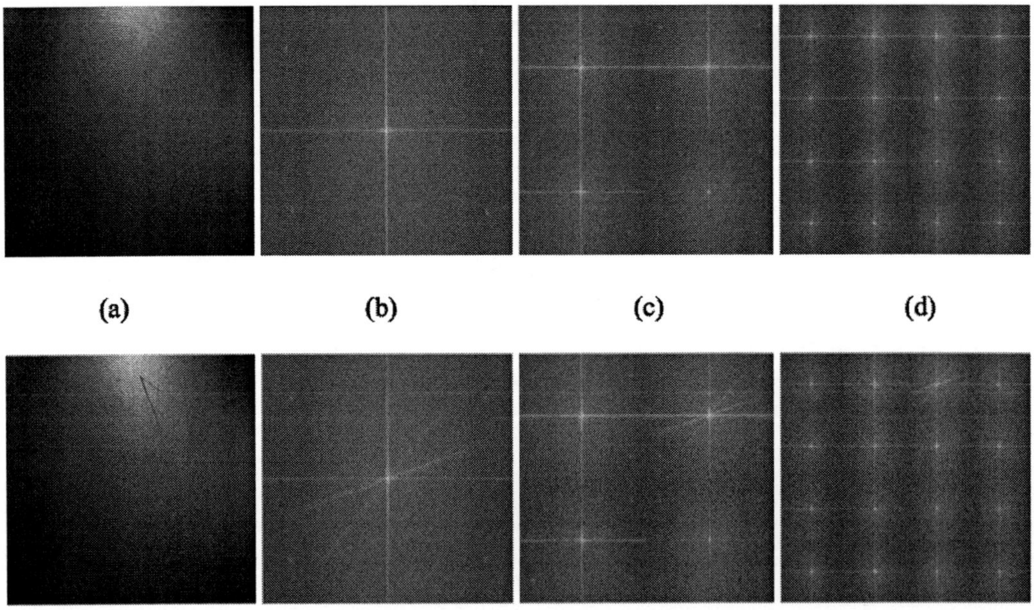

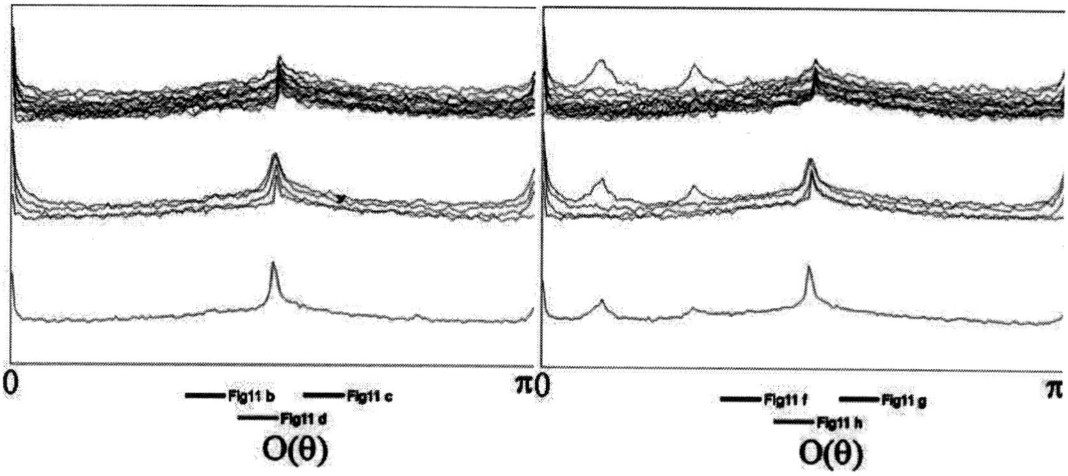

Figure 11. Texture invariance for several spatial resolutions.

The counting procedure then is to obtain the processed image and to calculate the value for its texture. If no defects are found, the next image is obtained. However, if a defective surface is detected, the image is subdivided and the DFFT calculated. The perfect zones are eliminated to focus only on the areas with defects, which must be subdivided again. The algorithm is stopped according to a user preset resolution value.

How to determine if there are several defects with the same direction? Figure 12 shows two examples, and the analysis is carried out as it has been explained so far. Again, several additional maxima appear, one per direction in which defects are found.

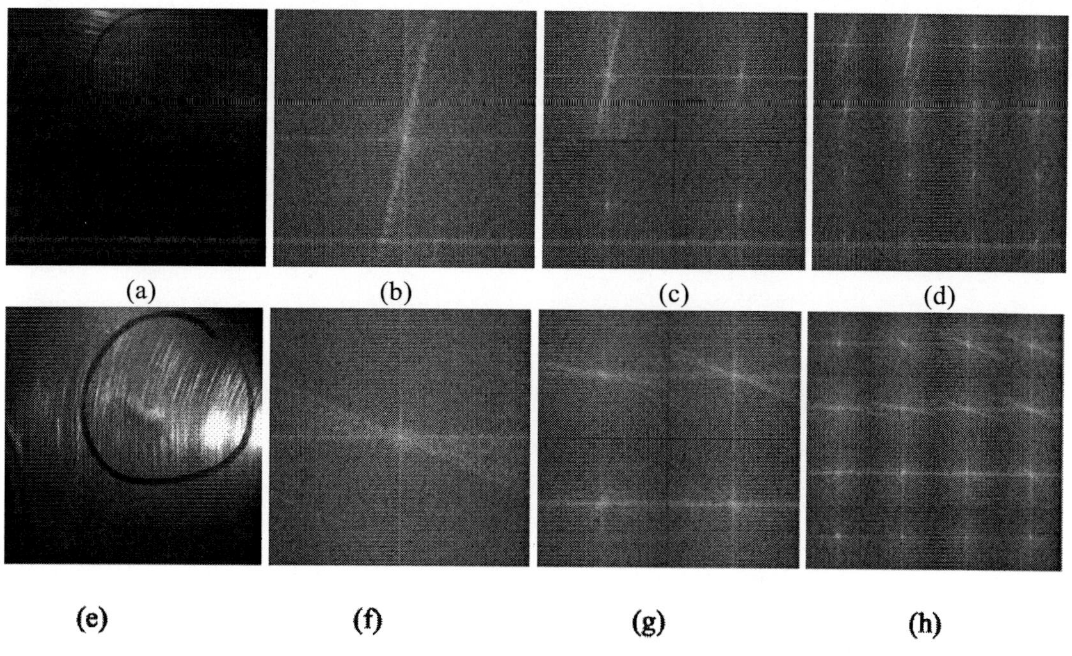

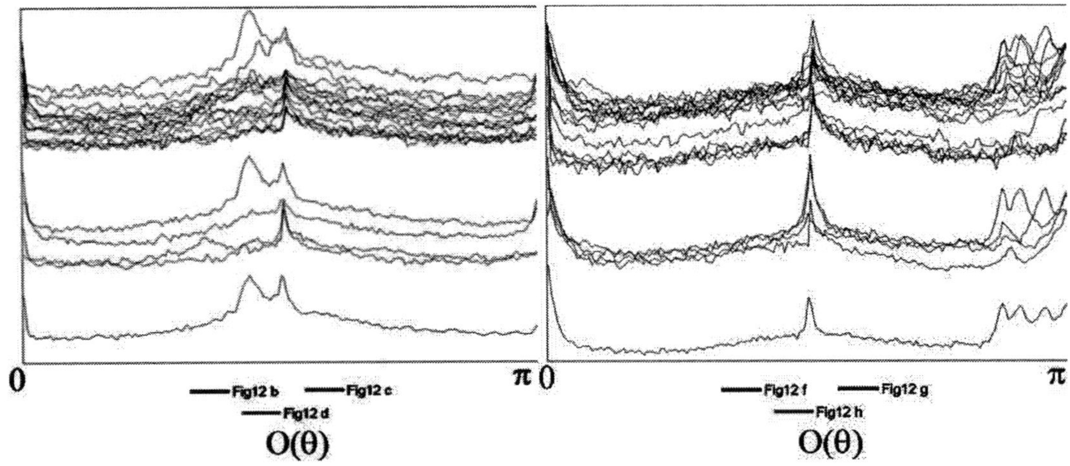

Figure 12. Detection of multiples defects in an image.

To know if there are one or several defects with the same direction, the modulus of the DFFT is analyzed in the direction of that local maximum. Figure 13 depicts the calculation of defects for images 10a, 11e, 12a. It can be seen that the shape is different, the two first cases are similar, whereas the third one shows a slower decay while having stronger variations.

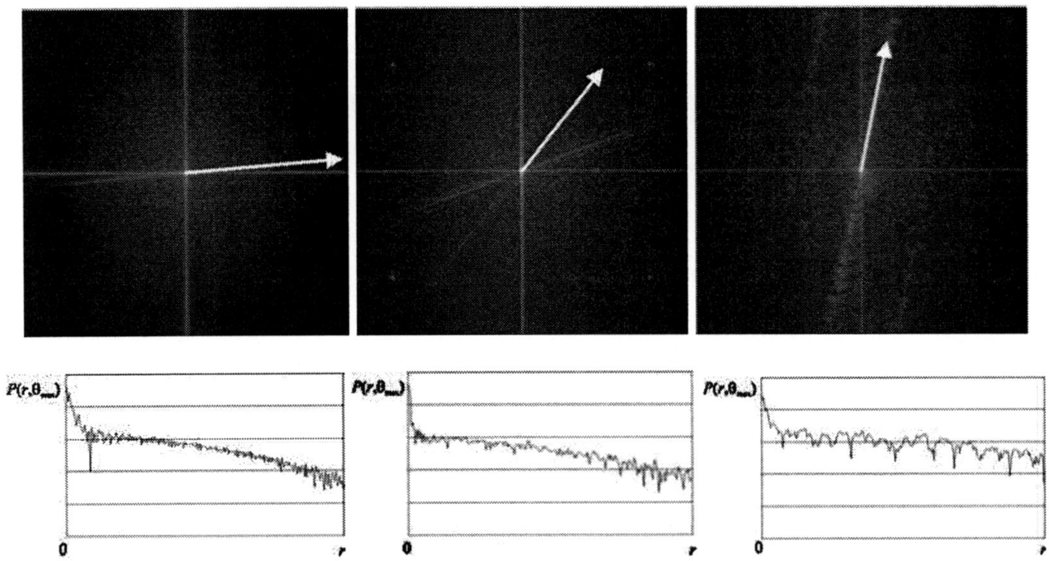

(a) (b) (c)

Figure 13. Analysis of the number of defects with the same direction.

An additional advantage of this procedure is that it can determine which zones of the image are saturated due to reflections. As figure 14 shows, for a certain resolution, a value for the modulus only exists for zero frequency, since for a saturated image, the pixel values are all at the maximum value, and the image is uniform.

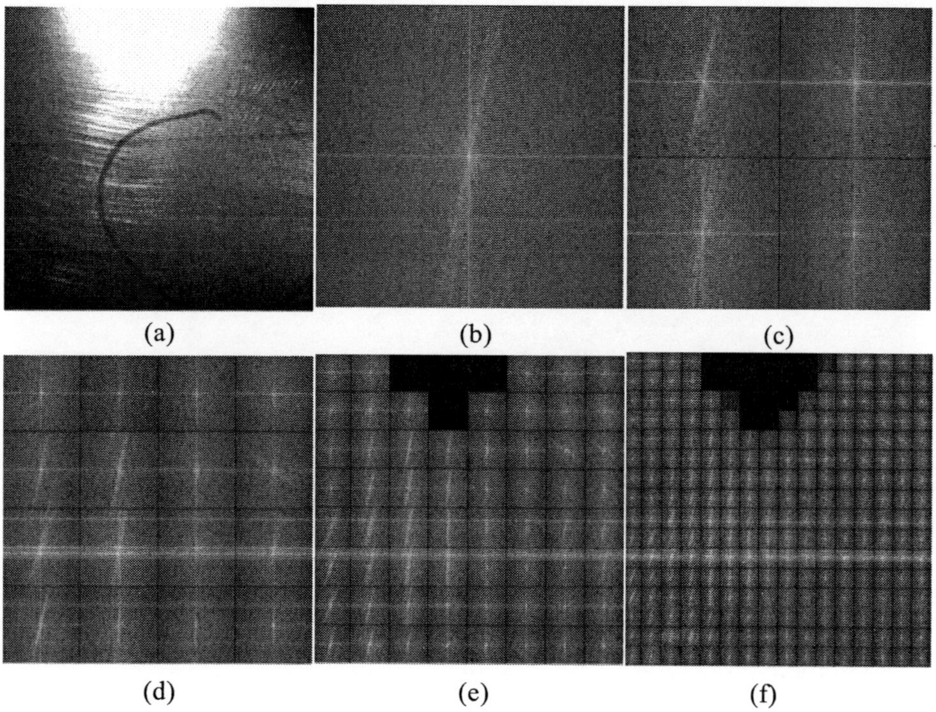

Figure 14. Determination of saturated areas within an image.

The inconvenience of this frequency analysis is its computational costs. That is why a combination of statistical pattern modeling and frequential texture analysis is proposed. With the first method, a quick selection of areas with nonconformities is achieved. The DFFT will then only be applied to these localized areas.

2.3 Experimental Results

The algorithm has been tested in a real situation. Several steel surfaces are tested to look for bumps and scratches during the manufacturing process of metallic pieces.

- The inspection system

The artificial vision system is composed of a progressive scan CCD camera which covers in each frame an area of 20x1 5 cm, and an illumination system of continuous current. The light gets to the surface via two optic fiber connecters which are located perpendicularly to each other in the same plane. The optic axis of the CCD camera is perpendicular to the plane of light, as shown in figure 15.

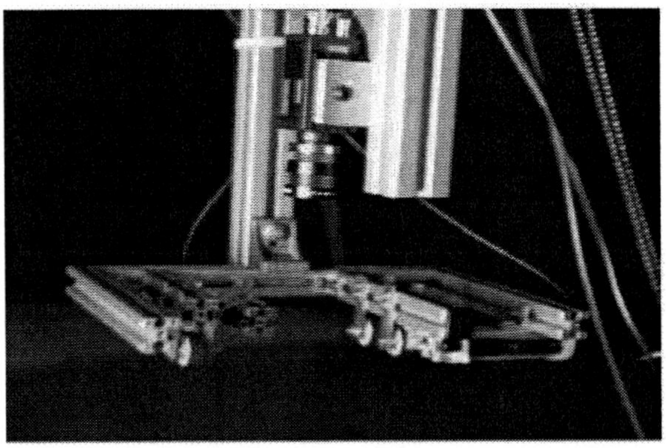

Figure 15. Artificial vision system.

The system has been implemented using an Asea IRB 2400 robot, to facilitate the inspection process. The inspection velocity of the robot is 200mm/s.

To calculate the quality confidence limits, twenty images are taken of flat surfaces without defects. Each image is composed of 768x572 pixels, so M = 439296 total pixels. A minimum value of 6 standard deviations is needed (k=6) to almost vanish the inspection errors. The test for normality has been successful for different series of twenty images.

- Detection of bumps and scratches

Figure 16 and figure 17 include the video image and the processed images of different defects that have been detected and classified in the metallic surface. Each example consists of six images. The first one is the image as seen by the video camera after applying the adjustment for illumination, the second one shows the processed image after performing the control limits check, the third one depicts the processed image after the filtering step, and finally, the last three show the texture analysis based on the Fourier transform, with different levels of splitting (1, 4 and 16 windows).

Bumps: The texture analysis when bumps are present does not show a dominant direction in the processing window that contains the defect. To correctly quantify the number of bumps, the number of windows must be increased. For the proposed algorithm, a resolution based in 16 windows has been implemented.

(a) (b) (c)

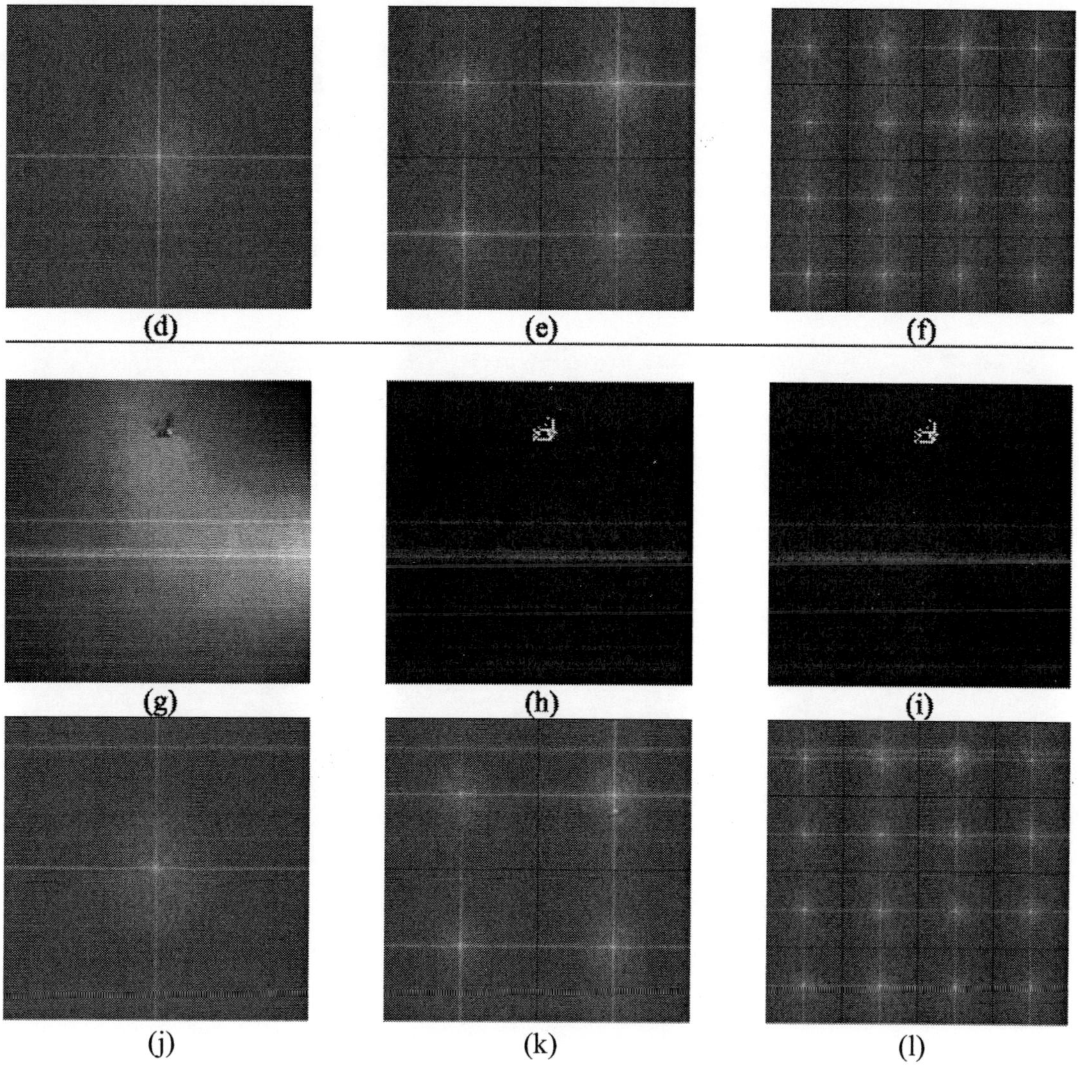

Figure 16. Processing of bumps.

Scratches: for this type of defect, a dominant direction appears when the texture analysis is performed. Again, a 16-window resolution is used to quantify the number of scratches.

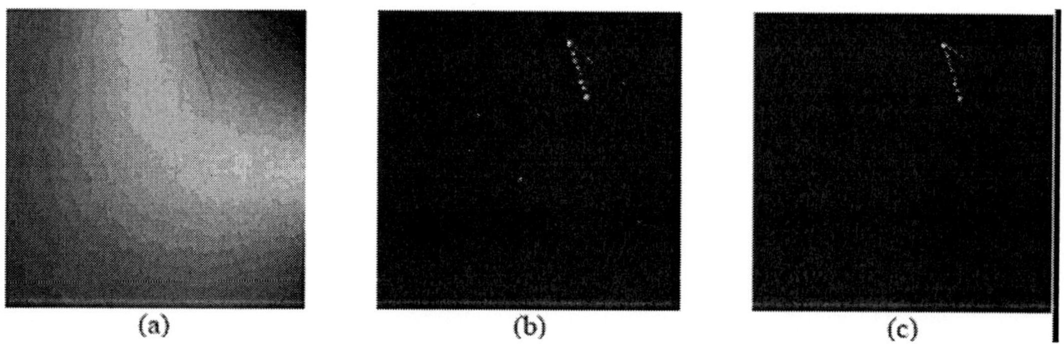

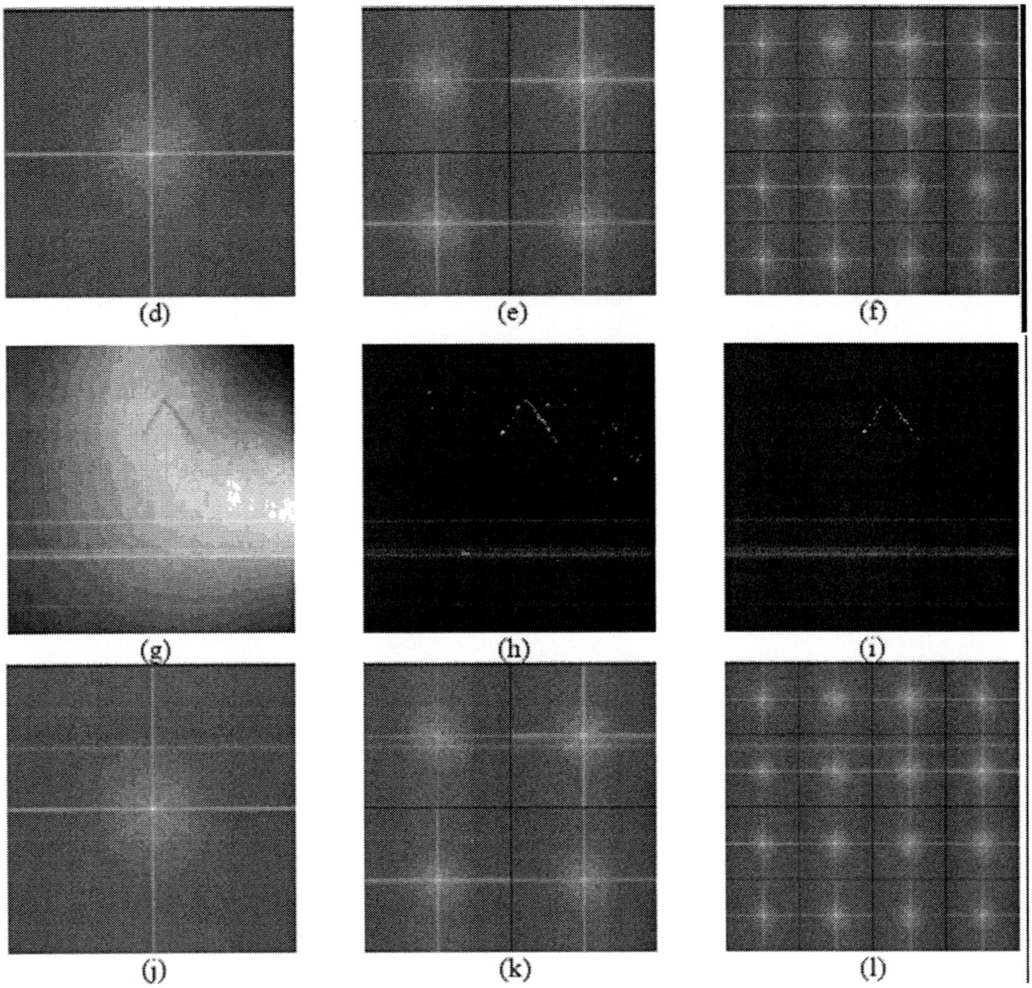

Figure 17. Processing of scratches.

The system has been so far designed to analyze surfaces with small curvature variations. For different curvatures, the predetermined patterns are different in the same surface, eliminating the possibility of controlling any type of surface. However, the proposed system can handle small variations, like the one presented in figure 18.

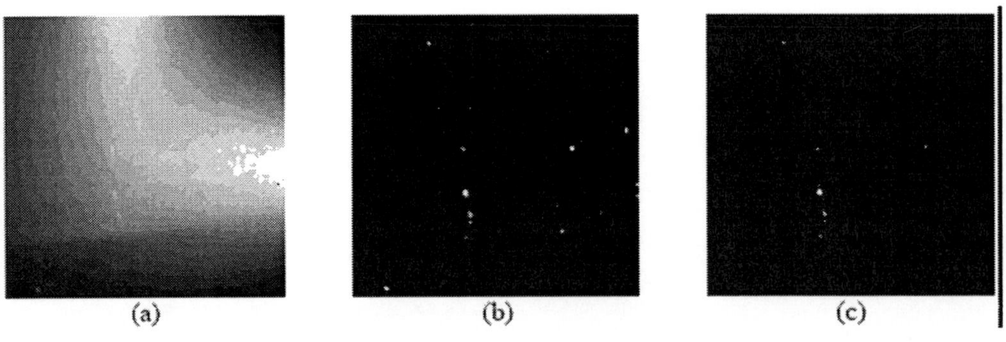

Figure 18. Images of curve images.

3 Production Control System

The last phase of the methodology is the control of the production line. The objective is to determine if the production process stays stable and in statistical control and, second, if it is capable of meeting quality requirements. Once each surface is scrutinize to determine if it is defective or not and the defects quantified and discriminated by type (first phase), the evolution of the number of each defect is monitored along time in a simple chart that it is intelligently updated electronically and visually controlled by the operators.

3.1 Total Quality Management

Management is responsible for meeting customers requirements. As a prerequisite, the production process must be in statistical control, which means that for a given set of inputs, the distribution of the output values is known. Then, it is possible to compare the natural limits of the process with the requirements.

What lies beneath the control of the line is then the understanding of the variation in the output values, specifically, the effect of normal causes of variation (like temperature, humidity...) and special causes (like operators, machines...). To improve the process, special causes must be removed and normal causes minimized to reduce the variability in the output.

In other words, for the total quality process to be worthy, management and operators should [19]:

- Recognize that variation is present and part of everything
- Understand something about the theory of variation and its causes: common versus special
- Use data appropriately so that variation can be recognized, interpreted and reacted to.
- Develop the strategy for dealing with both causes

The operators are then a crucial part of the control process. The sooner a fault is found in the line, the sooner the process can be stopped and improved. To fulfill their duties, operators must be provided with tools to [19]:

- Know whether the process is capable of meeting the requirements
- Know whether the process is meeting the requirements at any point in time
- Correct or adjust the process or its inputs when it is not meeting the requirements

One of the tools that is commonly used is charting, or the monitoring on line of the process using graphical techniques, like quality charts. These are graphs in which measurements are plotted along time and a set of specific rules determine if the process is in control or not. With the help of charts, not only the output distribution is monitored continuously when the process is in control, but also deviations in terms of mean behavior and variability are quickly detected. Once in control, it is also possible to calculate the degree in which the requirements are met.

What follows is a description of the proposed control charts and the capability indices, with an initial description of what is meant by variation.

3.2 Understanding Variation

In repetitive processes, no matter how well they are defined, not all of the outgoing products look exactly alike. There exists what is called *variation* in the final products.

The causes that provoke this variability within the processes might be classified in two types:

1. Special causes: They are the causes that produce certain anomalies in the products, which disappear as soon as the causes disappear. The result is an unstable process which produces inhomogeneous products. The process is said to be "out of statistical control".
2. Common causes: Even when all of the factors that have effect in the process are "in control", there are still another set of factors whose combined effect provoke small variations in the expected outcome. They are random in nature and they do not affect the stable behavior of the process along time.

3.2.1 Types of Processes

Regarding the causes of variation present, a manufacturing process may be classified as:

- Stable process: the one in which only non assignable sources of variation are found. The outcome of the process does not vary with time: mean behavior stays constant and so does the variation. As a consequence, the results are predictable.
- Unstable process: if only assignable causes are present. It is not possible to predict the outcome of the process.
- Capable process: A process is capable with respect to client specifications if the process output distribution ($\mu + 3\acute{o}$) lies within the requirements.
- Non-capable process: If requirements are not fulfilled, the process is non capable, and the assignable causes must be removed to reduce variability.

3.2.2 Methodology of Analysis

The objective of the quality control process is to detect any possible deviation that might cause a reduction in the quality standard with respect to client specifications. The goal is to determine if a given process is able to satisfy the clients both in terms of stability and capability.

The analysis requires that the following phases are consecutively executed, each with its corresponding statistical tools:

- *PHASE I: Stability:* the goal is to determine if the process is in statistical control. All the assignable causes must be eradicated.
- *PHASE I: Capability:* the goal is to determine if the process meets customers specifications. If not, the non assignable causes are to be found and removed.

For a process to be capable, it must be stable first. 3.3 Tools to measure stability: Control Charts.

The control charts help visualize the evolution of a parameter along time. If observations are plotted versus time, and then compared with the natural limits of the process, a Control Chart is obtained (Figure 19).

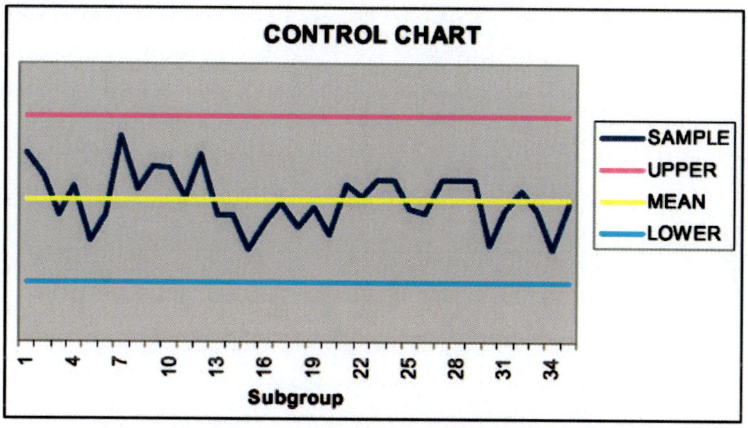

Figure 19. Developing a control chart.

As already mentioned, the statistical foundation to calculate these natural limits is that the outcome of the process follows a known distribution, usually the normal distribution.

The conditions under which each sample or observation is taken must be identical or homogeneous, so that known assignable causes are avoidable. The data represented in the chart correspond to individual products or group of products selected in such a way that each unit in a given time frame has the certainty of being produced under the same circumstances, so that the measure variability is due to non assignable causes. Then, any change between observations is due to assignable causes, subject to be analyzed.

3.3.1 Development

Therefore, three design parameters are crucial to develop a good control chart:

Sample size (n): number of units measured per subgroup.
Frequency (f): time between subgroups.
Number of samples (N): number of observations that must be taken to calculate the natural limits.

The control of the process quality is performed in three steps:

1. Calculation of the natural limits by taking N samples of size *n*.
2. Development of the control chart, which includes:
    - Process mean behavior, denominated CENTER LINE (CL).
    - Natural limits of variation, calculated so that 99.73% of the possible outcomes are included:

a) LOWER CONTROL LIMIT (LCL) $(\cong \bar{x} - 3s)$

b) UPPER CONTROL LIMIT (UCL) $(\cong \bar{x} + 3s)$

3. Implementation of on line control with the measurement and inclusion of new subgroups in the chart; for each new sample, a decision is made to determine if the process remains stable or an assignable cause has appeared.

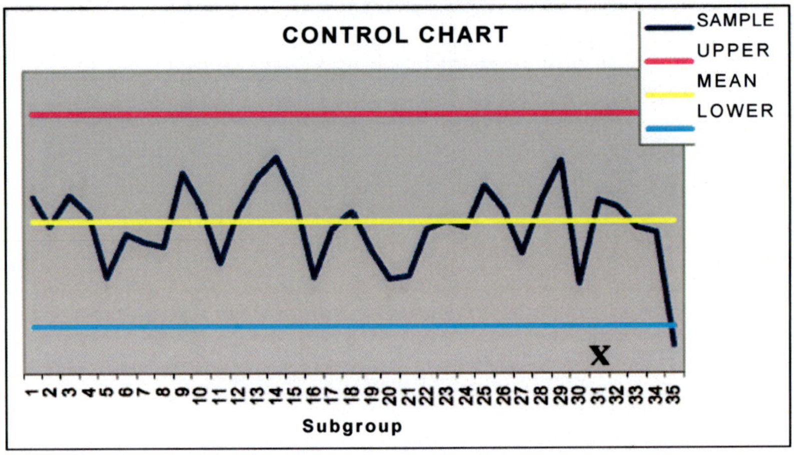

Figure 20. Updating a control chart.

The X (Figure 20) indicates an out of control signal. The process must be stopped and the problem found.

3.3.2 Types of Charts

There exist two different types, depending on the measure they include.

Variables Charts

The measure to be included is measurable in quantitative units. Each variable to be studied needs an independent chart. The parameter *n* is usually small (between 1 y 5), since sampling is usually costly *(example: shear strength of a rod).*

- • X /R charts: a pair of charts in which the mean behavior is studied with the X chart and the variability with the R chart (R for range). The values included in the X chart are the mean values of samples of *n* identical items. In the R chart, the ranges (maximum - minimum) are recorded.
- I/MR charts: a pair of charts in which the mean behavior is studied with the I chart (I for individuals) and the variability with the MR chart (MR for moving range). The values included in the I chart are direct values of a sample of the item, which is unique. In the MR chart, the ranges between consecutive samples are recorded.

Attributes Chart

The measures are numerous and some qualitative. All of them might be conjointly studied and *n* is usually large *(example: number of disconformities in a lot).*

- p charts: are used whenever the units produced in a lot of size *n* are classified as conforming or nonconforming. In this case, the underlying distribution of nonconformities per lot follows a binomial distribution, with an average proportion of *p* defective elements, and a proportion of $q = 1 - p$ conforming elements.
- c charts: they have been developed to control and analyze a single, very complex unit which might present several "non conformities".
- u charts: It is similar to the *c* chart, but more than one unit are analyzed simultaneously. This new chart is called the *u* chart, and it calculates the number of defects per sample *n*.

3.4 Tools to Measure Capability: Indices

When and only when the process is in control, it is the time to check if the process meets customer specifications. Following statistics theory, a set of indices have been developed to measure the existing relationship between the capacity of the process and the desires of its clients. Any index has the following structure:

$$\text{Index} = \frac{\text{Voice of the client}}{\text{Voice of the process}} \tag{9}$$

where the "voice of the process" measures the width of the process output distribution and the "voice of the client" measures the width of the desired specifications.

If the value of the index is 1, it seems clear that the process satisfies the requirements in almost all the cases, since the natural limits of the process coincide with the specifications:

Therefore, the processes with a capability index smaller than 1 will not be capable, not meeting the requirements in every case, whereas the ones with an index greater than 1 will be capable of almost always meeting the requirements.

The most commonly used indices are:

- Variables indices: Cp , Cpk, Cpm
- Attributes indices: e-ratio

3.5 On-line Control

Let's return then to the production system in hand. The visual inspection system is used to monitor the production of metallic surfaces, counting the number of defects per unit. The control chart to be used online is then the c chart for nonconformities.

What follows is, first, a more detailed definition of a c chart and its development so it exactly applies to the control of metallic surfaces. Second, the control rules that are used to detect assignable causes are presented. Third, the index for the quantification of the capability is introduced. Finally, the implemented charts in the production linc are listed.

3.5.1 Chart for Non-conformities

Known as c charts, they have been developed to control and analyze a single, very complex unit which might present several "non conformities" [10]. The following is a list of possible applications of the c chart:

- Number of nonconforming rivets in an aircraft wing or fuselage
- Number of surface imperfections observed in a galvanized sheet or a painted, plated or enameled surface of a given area
- Number of seeds (small air pockets) observed in a glass bottle
- Number of surface imperfections observed in a roll of coated paper or sheet of photographic film

Also, the underlying distribution of the number of defects follows a Poisson random variable, provided the following assumptions hold [12]:

- Each defect can be discretely counted.
- If the sample is subdivided such that only one defect could occur within a subdivision, then the probability of finding a defect within any subdivision is constant and small when the process is in control.
- This probability of finding a defect within a subdivision does not depend upon another subdivision having a defect.
- The sample consists of a very large number of such subdivisions.

Clearly, the number of defects per metallic surface is liable to be represented in a c chart. In order to develop the chart, let us define:

c_i = number of defects per inspected unit
N = number of inspected units

Then:

$$\bar{c} = \frac{\sum_{i=1}^{N} c_i}{N} \tag{10}$$

The c charts present the following limits if the normal approximation to the Poisson is used:

- Center line:

$$CL_c = c \tag{11}$$

- Upper control limit:

$$UCL_c = \bar{c} + 3\sqrt{\bar{c}} \tag{12}$$

Lower control limit:

$$LCL_c = \max(0, \bar{c} - 3\sqrt{\bar{c}}) \tag{13}$$

It is worth noting that the lower control limit usually takes a value of 0, the smallest possible value the variable number of defects can take. For that reason, the *c chart,* as depicted in Figure 21, does not present a symmetric structure, but one with the center line much closer to the lower rather than to the upper control limit.

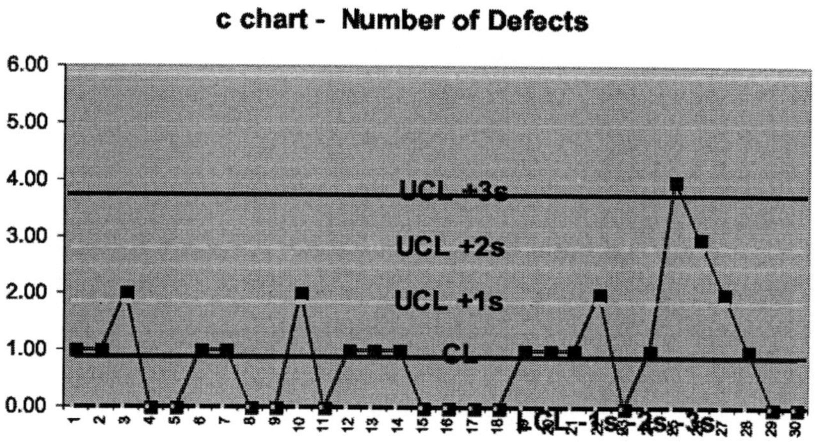

Figure 21. Example of c chart

However, this approximation to the normal distribution does not work too well in all situations, that is, it clearly depends on the value of $\bar{c}$ The smaller its value, the better the approximation. The problem that arise then is that the objective of the chart, which is to detect special sources of variation, is not achieved appropriately since the percentage of points outside the upper control limit might significantly vary from its desired value of 0.135% (percentage outside $\mu+3\sigma$ for a normal distribution).

For that reason, and to maintain the control philosophy, what is proposed following [10], is to situate the control limits to maintain the percentages corresponding to $\mu+k\sigma$ for the normal distribution:

	LCL	LCL	LCL	CL	UCL	UCL	UCL
Normal	0.135	2.275	15.866	50.000	84.134	97.725	99.865
k	-3	-2	-1	0	1	2	3

Knowing that the Poisson distribution with a mean of $\bar{c}$ has the following distribution:

$$f(x) = \frac{\bar{c}^{x} e^{-\bar{c}}}{x!} \quad \text{for x=0,1,2,...} \tag{14}$$

and therefore:

$$F(x) = \sum_{i=0}^{x} f(i) \tag{15}$$

it is possible then to easily determine the limits and the center line of the *c chart* according to the Poisson cumulative distribution function (Figure 22).

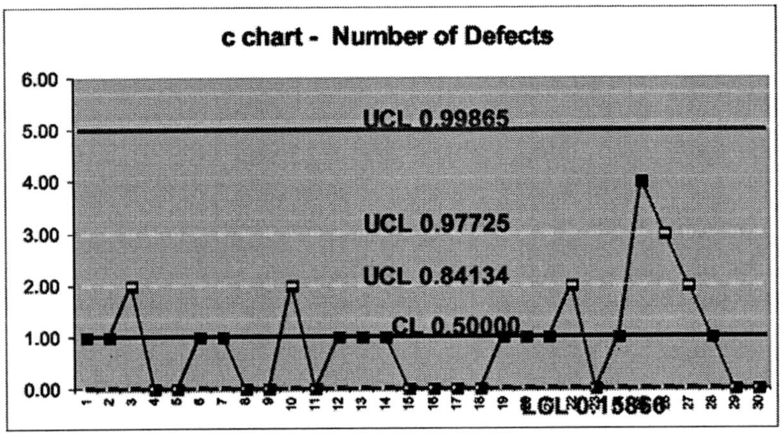

Figure 22. c chart with probability limits.

If the two last figures are compared, the difference between the two approaches is further explained. Whereas in the first figure, the normal approximation one, the process looks out of control, the process is really stable, as depicted by the more exact Poisson chart.

3.5.2 Updating the Chart and Controlling the Process

The chart is the tool then that is used to control the process. Each new observation is included in the chart and a set of rules are evaluated to determine if the process remains stable or for some reason it has gone out of control.

The rules to stop the process are the following [12]:

- One observation lies outside zone A
- Two observations out of three lying in zone B of the same side
- Four observations out of five lying in zone C of the same side
- Eight observations in a row lying in the same side of the center line
- Six observations in a row steadily decreasing or increasing
- Fifteen observations in a row lying within zone C
- Eight observations in a row lying outside zone C
- Fourteen observations in a row alternating up and down

where zone C is the one covering $\bar{x} \pm 1s$, zone B covering between $\bar{x} \pm 1s$ and $\bar{x}+2s$ and zone A covering between x +2s and UCL.

The first four rules apply to either side of the center line, whereas the last four relate to the whole chart. For symmetric charts it looks reasonable to implement the whole set of rules, whereas for asymmetric charts, like the attributes charts, only the first four are appropriate.

3.5.3 Measuring Capability

The most widely used index is the *e-Ratio*:

$$e - ratio = \frac{e_{max}}{\bar{e}} \tag{16}$$

where

e_{max} = maximum mean value allowed by the client of the statistic included in the chart (*p, u, c*)

$\bar{e}$ = average of the corresponding attributes statistic

The *e-Ratio* can take the following values:

e-Ratio < 1, if $e_{max} < \bar{e}$
e-Ratio = 1, when $e_{max} = \bar{e}$
e-Ratio > 1, for $e_{max} > \bar{e}$

The specific ratio for the *c chart* is:

$$c - Ratio = \frac{c_{max}}{\bar{c}} \tag{17}$$

If the ratio is lower than 1, the process is not capable, not satisfying the requirements of the client. The necessary improvement is to study and reduce the effect of the special causes of variation.

3.5.4 Proposed Charts

The actual production control system takes data from the visual inspection system. For each surface, the number of defects, in total and by type, are obtained and recorded in the following charts:

- c chart for nonconformities per unit
- c chart for bumps per unit
- c charts for scratches per unit

By splitting the chart in two, the out of control signal might be detected sooner, since the overall chart might include compensation between the two types of defects. Besides, since the whole set of charts is automatically updated, the hassle of introducing the data and controlling the chart is eliminated and a clearer picture of the performance of the process is obtained.

4 Conclusion

Machine vision provides innovative solutions in the direction of industrial automation. The combined use of techniques from different fields like computer vision (spatial and frequential processing), statistical theory (normality tests, confidence intervals) and production control (quality control charts, capability indices) has proven to be successful in a real situation.

An intelligent visual inspection system has been developed and applied to the control of a continuous production line of metallic surfaces. The system is divided into the identification and classification of defects on each metallic surface and the control of the production line.

The computer vision system is able to quickly detect defects on each individual surface by the correct characterization of its visual aspect. The detection method is a hybrid between the two usual methods of pattern comparison and theoretical decision. The classification and quantification method is based on a texture analysis algorithm which uses the Fourier transform.

Besides, by continuously updating quality control charts for nonconformities, the production is monitored over time to detect out of control situations. The charts for nonconformities receive data from the visual inspection system and then control rules are used to determine if the production line stays stable or it has gone out of control due to special causes.

References

[1] Bahlmann, C., Heidemann, G., Ritter, H. (1999). Artificial neural networks for automated quality control of textile seams. *Pattern Recognition,* vol. 32, pp. 1049-1060.

[2] Caelly, T. (1994). *A brief overview of texture processing in machine vision.* Early Vision and Beyond, pp. 79-86. MIT Press.

[3] Chung, Y.K., Kim, K.H., Kim, H.B. (1997). Extracting of line features for automatic inspection of vehicle doors. *International Conference on Signal Processing Applications and Technology,* San Diego, California.

[4] Cootes, T.F., Page, G.J., Jackson, C.B., Taylor. C.J. (1996). Statistical grey-level models for objects location and identification. *Image and Video Computing,* vol. 14, pp 533-540.

[5] D'Agostino, R.B., Stevens M.A.: 1986. *Goodness-of-Fit Techniques,* Marcel Dekker.

[6] Dodgson, N.A., Wiseman, N.E.: 1995. Graphic Design for Ceramic Tiles. *Eurographics.*

[7] Fernández, J., Fernández, C., Aracil, R.: 1997. On-line integral quality control of ceramics tiles by artificial vision. *Proceedings of the International Conference on Quality Control by Artificial Vision,* vol. 1, pp 156-161, Le Creusot, France.

[8] Ferreira, M.: 1998. Automatic Visual Inspection on Shoe and Textile Industries. *Proceedings of IASTED Conference on Computer Graphics and Imaging.* Halifax, Nova Scotia, Canada.

[9] Gonzalez, R., Woods, R. 2002: Digital Image Processing. *Prentice Hall.*

[10] Grant, E.L., Leavenworth, R.S.: 1988. *Statistical Quality Control.* McGraw-Hill.

[11] Hirvonen, J.: 2000. Measurement if the quality of mechanical pulp by machine vision. *Machine Vision News,* vol.5.

[12] Joglekar, A.M.: 2003. Statistical Methods for Six Sigma. *John Wiley & Sons: New Jersey.*

[13] Khashman, A.: 2000. Neural System for Monitoring Non-Linear Foetal Measurements. *2000 IEEE Nordic Signal Processing Symposium.*

[14] Kennedy, S.W., Hoffman, E.G.: 1987. Inspection and Ganging. *Industrial Press,* New-York.

[15] Kim, K.H., Kim, Y.W., Suh, S.W.: 1998. Automatic visual inspections system to detect wrongly attached components. *International Conference on Signal Processing Applications and Technology,* Toronto, Canada.

[16] Nelson, L.S.:1982. Control Charts for Individual Measurements. *Journal of Quality Technology,* Vol. 14, No. 3, pp. 172-174.

[17] Niemi, A.: 2000. Machine vision based control of the paper machine solution of a 200 years old problem. *Machine Vision News,* vol. 5.

[18] Noble, A., Nguye, V., Marinos, C., Mundy, J.: 1992. Template guided visual inspection. *Proceedings of the 2nd European Conference on Computer Vision,* pp. 893-901.

[19] Oakland, J.S.: 2003. Statistical Process Control. *Butterworth-Heinemann, Oxford, 5th ed.*

[20] Parker, J.R.: 1997. Algorithms for image processing and computer vision. Wiley Computer Publishing,.

[21] Randen, T., Husoy, J.H. 1999: Filtering for Texture Classification: a Comparative Study. *IEEE Transactions on Pattern Analysis and Machine Intelligence,* Vol. 21 Issue: 4.

[22] Rioux, M., Goldin, G., Blais, F.: 1993. Differential inspection of shapes using 3D optical measurements. *Proceedings of Optical 3D measurement techniques I: Applications in Robotics. Quality control and Robotics,* SPIE vol. 2252, pp.493-50 1.

[23] Silven, 0., Kauppinen, H.: 1996. Recent developments in wood inspection. *International Journal of Pattern Recognition and Artificial Intelligence,* vol. 10 nº 1, pp. 83-95.

[24] Smith, B.: 1993. Making war on defects: six-sigma design. *IEEE Spectrum,* vol. 30, N°9, pp. 43-47.

[25] Tantaswadi, P., Vilainatre, J., Tamaree, N. and Viraivan, P.: 1999. Machine Vision for Automated Visual Inspection of Cotton Quality in Textile Industries Using Color Isodiscrimination Contour. *Computer and Industrial Engineering.,* Vol. 37, June 1999, pp. 347-350.

[26] Veistinen, M., Simomaa, K.: 1999. Machine vision reveals defective magnets and qualify parts for assembling. *Machine Vision News,* vol. 4.

[27] Winchell, W.: 1996. Inspection and measurement in manufacturing. *Society of Manufacturing Engineers.*

In: Frontiers in Robotics Research
Editor: Max A. Denket, pp. 35-66
ISBN: 1-60021-097-X

Chapter 2

DEXTEROUS UNDERWATER INSPECTION MANIPULATORS: DESIGN CHALLENGES AND DEVELOPMENTAL ISSUES

Gerald Seet, Asokan Thondiyath, Michael Lau and Eicher Low
Robotics Research Centre
School of Mechanical and Aerospace Engineering
50 Nanyang Technological University, Singapore 639798

Abstract

When conditions permit, experienced divers are invariably deployed to inspect underwater installations for damages to structure and weld joints. With advances in Underwater Robotic Vehicles, unmanned inspection techniques promise to be a more cost effective and safe option. One of the major challenges to underwater robotic inspection is in the design of dexterous manipulators to operate in the underwater environment. The work environment can be highly constrained due to limited accessibility of the vehicle to the interconnected structures and the limited dexterity of existing systems. The design of special purpose, custom-made manipulators for such specialized applications is a challenging task. We present and discuss the design of a special purpose, custom made manipulator for inspection of underwater structures. A review of the state-of-art underwater manipulators is included. The functional requirements of an underwater manipulator for subsea inspection are discussed, and the desired performance requirements are identified.

The design challenges and developmental issues of a special purpose manipulator for underwater inspection are elaborated. A novel kinematic structure is proposed for Underwater Robotic Vehicle (URV) operation, having a 2 d.o.f launching stage and a 5 d.o.f inspection stage. In order to reap the benefits of both the hydraulic and electric actuators, a hybrid actuator-power structure has been adopted for the manipulator. Large torque and relatively lower positioning requirements of the launching stage dictate the use of high torque hydraulic actuators, for this stage. Similarly, relatively lower torque and high accuracy requirements of the forward inspection stage make the specifying of electrical actuators the obvious choice.

In underwater applications, reliability is a paramount consideration. Proportional electro-hydraulic servo valves are susceptible to particular contaminants in the hydraulic lines. The simpler poppet valve designs are more tolerant to such contaminants, but are used in a simple

full-on or full-off ('bang-bang') arrangement. This lacks the opportunity for fine motion control. In our design, the actuating current to the poppet valve solenoid is 'pulse width modulated' to allow a more precise control of the hydraulic actuator. This is achieved through more sophisticated switching of control signals by the embedded micro controller. Another major source of failure for underwater systems is the connectors. The novelty of design extends to the inspection section by the integration of intelligent actuator modules. This provides for a simpler mechanical and electrical interconnection configuration, resulting in a reduction in cable lengths and number of connectors. Underwater connectors remain a major cause of equipment malfunction.
Design philosophy, system architecture and design constraints are presented and discussed. Results of simulation studies, Experimental analysis, and optimisation problems are also presented. The article serves as a technical research paper and provides insights into a detailed case study of the processes and considerations of realizing a novel special-purpose manipulator for underwater robotic vehicle (URV) applications.

1.0 Introduction

Man's dependence on the ocean for food, transportation, natural resources, and security dates back to early modern civilization. With the growth of the shipping and oil industry, many tasks such as maintenance and repair of ships, fabrication and repair of sub sea structures, recovery or salvage of sunken objects, etc., necessitated underwater operations beyond the reach of human divers. High cost and low efficiency of human divers, as well as the lack of trained divers to perform complicated tasks in an extremely hostile environment, have led to the development of unmanned underwater robotic vehicles. These are used widely for sub sea operations worldwide. In the 1970s, the development of underwater robotics was boosted by rapid advancements in computer technologies. Advanced technologies such as new manufacturing techniques, new materials, advanced control algorithm, and locating and sensing systems have also helped the growth of underwater robotics. Underwater inspections, submarine installation, recovering shipwrecks, mining, military and scientific survey, etc., and the wide range of other applications, boosted the development and popularization of underwater robots further. In 1974, there were about 20 underwater robots operating around the world. By the year 1978, the number grew to 100. In 1994, there was an estimated 2000 underwater robotic vehicles of over 150 different types by over 60 manufacturers operating in various corners of the ocean. Their numbers are growing at a steady rate. Figure 1 shows the URV developed at the Robotic Research Centre at the Nanyang Technological University, Singapore, for inspection of submerged pipelines.

Underwater manipulators were developed to enhance submarine operations since manned submersibles experienced rapid growth in the 1950s and 1960s. They were primarily used for relatively simple operations such as picking up sunken objects or ground samples. One of the examples was the manipulator, which was installed for the "Bathyscaphe Trieste", the research and exploration vessel built in 1953 and used by US Navy for various missions. With advances in robotic and manipulation technology, they are being used for increasingly complex tasks like cleaning, inspection and repair of subsea structures.

Figure 1 Underwater Robotic Vehicle Developed at NTU

Much research in the area of underwater robotics has been directed at reducing the need for human intervention, through increasing autonomy of the vehicle and subsystems. Along with advancements in underwater robotic vehicles, unmanned inspection techniques are being developed. With the tremendous growth in sub sea communication cables, oil and gas pipelines, and erection of ocean structures, subsea inspection has become a "hot field" in underwater robotics. Development of sub systems for URV to assist unmanned subsea inspection is crucial in tapping this potential.

2.0 Applications and Deployment

2.1 Subsea Inspection

Underwater structures require periodic inspection and maintenance to ensure a safe working environment. Unmanned underwater inspection techniques are being developed to replace the human divers in carrying out hazardous and complicated inspection tasks. A typical scenario of an unmanned inspection is shown in Fig.2. A URV with an onboard inspection system is deployed from the base station and remotely navigated to the targeted workspace by the pilot (URV operator), located at the base vessel. Using mechanical constraints, the URV is docked to the structure and inspection is carried out. Depending on the autonomy of the subsystem, operations are performed with differing levels of pilot intervention.

Non-contact type NDT (Non Destructive Testing) methods are generally employed for the inspection of underwater structures for crack identification. The ACFM (Alternating Current Field Measurement) is a commercially available NDT technique widely used in sub sea inspection. The ACFM technique is a non-contacting electromagnetic technique for the detection of surface breaking defects in conducting materials. The ACFM probe induces a uniform electric current into the material to be inspected, which then produces a magnetic field, which will be disturbed around the edges of a defect. Small detectors or sensors are built into probes, which are used to detect these magnetic field disturbances. Two components of the magnetic field are measured, Bx and Bz, the former to estimate crack depth and the latter to estimate crack length. These measurements together with software algorithms are used to determine the accurate length and depth of the defect. This method is suitable for electrically conductive materials including steel, stainless steel, aluminium, titanium and carbon re-enforced plastic materials. Case studies where ACFM has been used to size defects can be found in [1]. ACFM gauges are commercially available for interfacing with underwater robotic vehicles. The probe is required to be held firmly by the onboard arm gripper, and manipulated along the inspection surface.

The successful completion of the inspection depends on many factors, namely:

- Docking position of URV
- Manipulator kinematics, dexterity, type of actuation, etc.
- Work environment
- Inspection method adopted

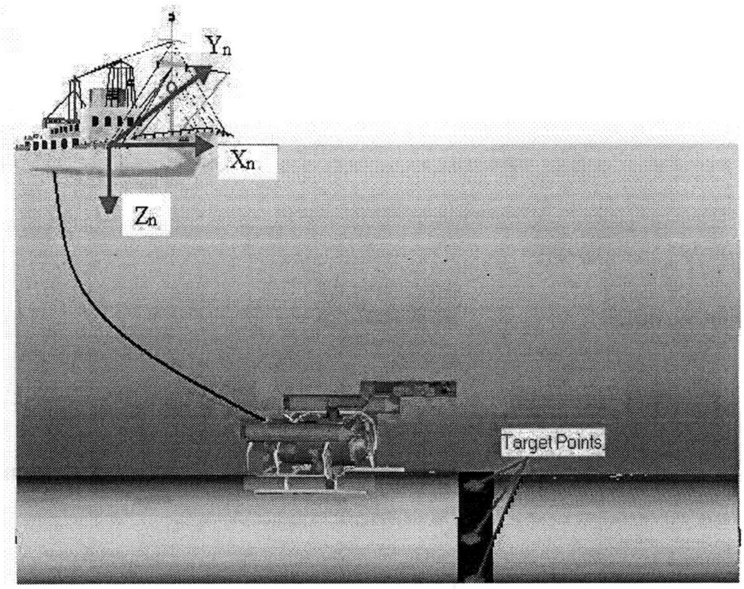

Figure 2 Typical Scenario Of Subsea Inspection Using URV

The size, weight, kinematic structure, and folded configuration are significant factors in deployment. Prior knowledge of the type of application, proper selection of manipulator configuration and actuation methods, based on the intended application, will eventually help to improve the efficiency of the manipulator in a real time inspection application.

2.2 ROV Intervention and Inspection Manipulators

Along with the development of Remotely Operated Vehicle (ROV) technology, the need for special purpose manipulators to perform dexterous underwater intervention tasks was also identified. Research and developmental activities in this area has also been receiving renewed interest [2,3,4]. Over the years, many commercial manipulators have been built and used in sub sea tasks.

Underwater manipulators are generally anthropomorphic, which depicts a human arm. Grippers or other changeable tools are attached at the end effectors. Unlike their terrestrial counterparts, most underwater manipulators consist of only revolute joints. They generally have four or five degrees of freedom. It is considered sufficient to allow them to position and orientate tools properly for their tasks. A manipulator with more than five degrees of freedom is rare, but additional degree of freedom increases its overall dexterity. Manipulators are used for a variety of tasks, these include:

- Cleaning of welds for inspection
- Close visual inspection and Non Destructive Testing (NDT)
- Opening and closing of valves
- Salvage of sunken objects
- Cable laying and repair
- Clearing debris and fishing nets

Most of the existing underwater manipulators are hydraulically powered. Hydraulic drives offer several advantages over electric actuation system, such as higher torque to weight ratio, overload resistance and generally are more resistant to the corrosive marine environment. Rigid pipe or flexible hoses are used to convey pressurised hydraulic fluid from reservoir to actuators.

The lifting capability of manipulators differs greatly depending on the tasks they are to perform. For example, the manipulator Norson, which was developed by Norson Power Ltd., has a lifting capacity of 160N while Titan 3, developed by Schilling, can lift objects weigh up to 1100N. Lift capability of 250N – 1000N are common for most underwater manipulators. Perry-Slingsby's TA60 manipulator has a lift capacity of 3000N throughout its envelope and 3250N with its extended boom fully retracted. It has a wrist torque output of 300Nm. The permitted arm extension of an underwater manipulator, also, depends on the stability of the platform (the underwater robotic vehicle). If the platform is stable, such as in a docked position, fixed to the ground or structure, the underwater manipulator can be as long as 2.5m or more. If the platform were positioned dynamically in the water, a working space of that size, would be very difficult to achieve.

Inspection manipulators are widely used in NDT based inspection of weld joints. They carry the subsystems for accurate and reliable recording of the location of defects with respect to pre-established datums. The number of degrees of freedom (d.o.f.), reach, dexterity, and operating environment are some of the factors that influence the manipulator performance. Perry-Slingsby's TA series manipulators are widely used in inspection tasks. General Robotics (UK) has developed the ARM (Advanced Robotic Manipulator) system that offers a comprehensive package for underwater inspection and repair of weld structures.

3.0 Dexterous Inspection Manipulator: Design Issues

The design of underwater manipulators is significantly different from the design of factory floor mounted manipulators. Due to the differences in functional requirements of these applications, customized manipulators are necessary. The need for custom made, highly dexterous, multifunctional, manipulator for inspection and repair of submerged structures was identified during the conceptualization of the underwater project at NTU. The basic tasks to be performed by the manipulator were broadly identified as cleaning, inspection and repair of faulty weld joint of submerged structures. None of the commercially available manipulators satisfied the multifunctional requirements of the inspection manipulator. This was attributed to the demanding manipulator requirements, including size, envelope, lift capacity, joint mobility, kinematics and actuation methods. A custom designed manipulator was the only option.

The underwater environment is highly constrained and in general, very difficult to have a clear, free access to the submerged installations for inspection/repair tasks. The small sized ROV; presently being developed by NTU is expected to address this issue to a great extent by providing better access to the submerged structures. However, it was identified that in the absence of undesirable repeated docking, an underwater manipulator would still need a reach of 2 to 2.5 meters to carry out meaningful operations. This means a very large workspace and in turn, long link lengths for the manipulator. To carry out an inspection/welding task by the manipulator, it is necessary that the manipulator possess high dexterity and manipulability. The large link lengths of the manipulator, often leads to a loss of dexterity and manipulability of the manipulator. Therefore, one of the major constraints in the design of an under water manipulator is the compromise between the large workspace requirements and the manipulator dexterity. A large workspace for the manipulator is required also to ensure minimum number of dockings of the underwater vehicle, to cover the maximum work area.

Another constraint in the design is the folded configuration requirements of the manipulator. As the manipulator is mounted on a moving vehicle, the size, weight, kinematic structure, folded configuration etc. plays an important role in the dynamics of the vehicle as well as to the performance of the manipulator. The manipulator has to be designed in such a way that, it has a folded configuration, which add little disturbance to the free flight of the ROV. Based on the above constraints, the following design objectives/criteria have been defined as a framework for the design of the manipulator.

- The manipulator must be capable of tracking a complete weld joint by docking the ROV at one or two locations in the neighborhood of the joint. The number of docked positions for a single inspection task is to be kept to a minimum.
- It must be possible to carry out the inspection process by docking the ROV at a distance sufficiently away from the inspection point to have an unobstructed workspace for the manipulator to manipulate in. (Manipulator must have a very long reach).
- The manipulation space in an underwater environment is highly constrained. It is, typically, desirable to have the dexterous workspace towards the boundary of the actual workspace.
- From the practical design considerations and the available data regarding underwater structures, the total length of the manipulator had to be within 2 to 2.5 meters. This has been fixed on the basis of the available information on underwater structures and ROV capabilities [5,6]
- The kinematic structure must allow a folded configuration, which will ensure that the CG of the folded configuration is close to the CG of the URV and have very little influence on the dynamics of the URV during its free navigation.

- It is desirable that, the manipulator should permit the tracking of the weld joint with minimum movements of the lower joints, without compromising any other aspects of the inspection process.

Another important aspect of design is the type of actuation. There are three main methods for actuation, namely, electric, hydraulic, and pneumatic. These methods have distinctive characteristics, and relative advantages and disadvantages. Choosing a particular type of actuation depends on the torque requirements, speed of actuation, accuracy etc. Pneumatic actuation is used for low torque applications and is not suitable for an underwater environment. Some of the merits and demerits of choosing hydraulic or electric actuation for the manipulator is discussed below.

3.1 Hydraulic Actuation

Compared with mechanical power transmission methods, hydraulic systems are much more economic, practical and effective, especially when the power is to be transmitted to a moderate distance (several meters or above). Hydraulic power is transmitted by means of pressurized fluid in flexible (or rigid) pipes. Compared to a mechanical system which relies on gears, pulleys and levels; hydraulic system is less restricted in terms of the geometry of machines.

In comparison with electrical system, hydraulics can transmit significantly larger power. Hydraulic systems, in particular hydrostatic systems, rely on high-pressure fluid at low flow rates to transmit power. This minimizes fluid flow losses. In an electromagnet or electro-mechanical system, the power transmitted is limited by the saturation limit of steel in the presence of heating effects and power losses during transmission. Hydraulic actuators possess the highest power-to-weight ratio when compared to other means of power transmission systems [7]. Another advantage of hydraulic system is the compactness and ease of maintenance of the system. Hydraulic systems are generally controlled using a variety of valves. Hydraulic valves have relatively few moving parts, as compared to belts and pulleys in mechanical systems. A hydraulic system is capable of producing an output force much larger than the input force, without resorting to mechanical components such as gear and levers. Moreover, hydraulic fluids in a hydraulic system can help dissipate excess heat away from the system and can act as a lubricant to protect moving parts from wear and tear. All these contribute to a safe, simple but reliable system, which requires relatively, lower supervision and maintenance. Hydraulic systems also enjoy other advantages such as it is a reversible system, possess inbuilt protection against overload, is easy to employ, and capable of variable speed control.

Hydraulic systems, however, have some drawbacks. Hydraulic power is not as readily available as electrical power. In order to operate, a hydraulic system needs a hydraulic pump, pipes, valves, filters, regulators and the actuator. In comparison, a simple electric system only requires a battery, wire conductors, switches and motor. The components required by electric system are common and can be obtained easier than components for a hydraulic system.

The hydraulic fluid is one of the most important components of a hydraulic system, and the components that require the most attention. Generally, hydraulic oils are used. It is always messy and the leakage problem is almost impossible to solve. Improper sealing in pipes and hydraulic casing can cause oil leakage problems, which not only results in a degradation of performance, but also produces unpleasant smell and can be highly flammable. This imposes an upper temperature limit in which a hydraulic system can operate. Proper care needs to be taken to protect the hydraulic fluid from contaminations. Contaminants in hydraulic oil tend to clog valves and pumps, resulting in a system failure.

In order to increase the performance and reliability, components of hydraulic systems are manufactured under high quality control. To withstand the intense pressure during operation, the valve, for instance, must be manufactured with high quality materials and to high standards. Tight tolerance control is also imposed to ensure proper clearance between mating parts to reduce oil leakage. Therefore, hydraulic systems are always expensive and have lower potential for economies of scale. The need for a fluid reservoir and the need to contain high pressure, generally results is a bulky system. The power source can, invariably, be separated from the control and actuators.

3.1.1 Hydraulic Actuators

While hydraulic pumps convert mechanical power into hydraulic power, hydraulic actuators convert hydraulic power to mechanical power. Actuators and pumps can be regarded as counter parts of each other. The operation principle is the same but reversed. Pumps add energy into fluid for transmission to the point where it is required. Actuators extract the energy from fluid and do the work.

There are two major categories of hydraulic actuators, namely:

i.) Linear hydraulic actuators
ii.) Rotary hydraulic actuators

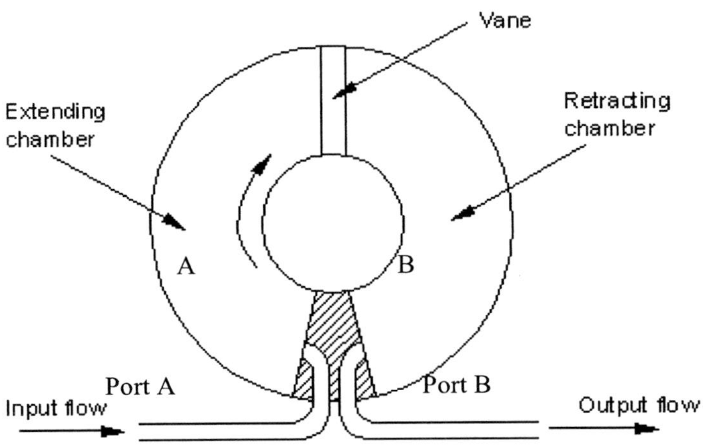

Figure 3 Vane Type Hydraulic Actuator

Linear hydraulic actuator provides linear motion when it extends and retracts. It is the simplest type of hydraulic actuation. As opposed to linear actuator, a rotary actuator provides rotary motion. The working principle of rotary actuator is similar to that of linear actuator, and is depicted in Fig.3. Pressurized fluid enters the actuator at port A. It will increase the pressure at the chamber A, connected to port A. The pressure difference between two sides of the vane results in a net torque, which turns the rotor and performs works. The rotation of rotor will increase the pressure in chamber B, causing the fluid to flow out of the chamber via port B. In this case, chamber A and B is called the extending and retracting chamber, respectively. To reverse the rotation of actuator, simply reverse the direction of flow. Due to the existence of the stationary abutment, the rotation of rotary actuator is limited. Another variation of actuator has two vanes, reducing the rotation range into half. However, in this variation, the torque-carrying capacity is twice that obtained by a single vane design.

3.1.2 Hydraulic Control Strategies: On-Off Control

There are different control strategies employed in hydraulic control applications. On-Off control, proportional control and servo control are the most common. Control of hydraulic systems with on-off solenoid valves is the simplest of the electro-hydraulic control scheme. The valves have two states, either fully open or fully closed. Fluid is allowed to flow pass the valves when they are opened, and the flow stops when the valves are closed. Hence, opening and closing the valve regulates the flow of fluid. This is characterized by low cost and relatively limited accuracy. However, it has been used successfully for almost a century with relatively few changes to the original principles.

In an On-Off control hydraulic system, the solenoid valves are normally controlled by the programmable logic controller (PLC). In one of the most basic designs, the PLC sends signal to open a solenoid valve, and fluid is allowed to flow into a cylinder and extend it. When the intended displacement is reached, a follower on the cylinder may trigger a limit switch, which sends a signal back to the PLC. Upon receiving this signal, the PLC will close the solenoid valve, stopping the flow and movement of the cylinder. The solenoid valves in On-Off control circuits may be arranged with various transducers and logic circuit to achieve various control targets. They are usually employed for displacement control and flow direction control. They are seldom extended to further controls such as velocities and acceleration controls. The construction of a solenoid valve is relatively simple, and these valves are usually deployed for the control of relatively small flow rate. Therefore, the force capacity of the solenoid is always limited to less than 65N. To handle larger hydraulic system, pilot operated solenoid valves are always employed.

Flow control valves are employed to regulate the flow rate of a hydraulic circuit, and thus the speed of a hydraulic actuator. The simplest type of flow control valve is a poppet type on/off poppet valve.

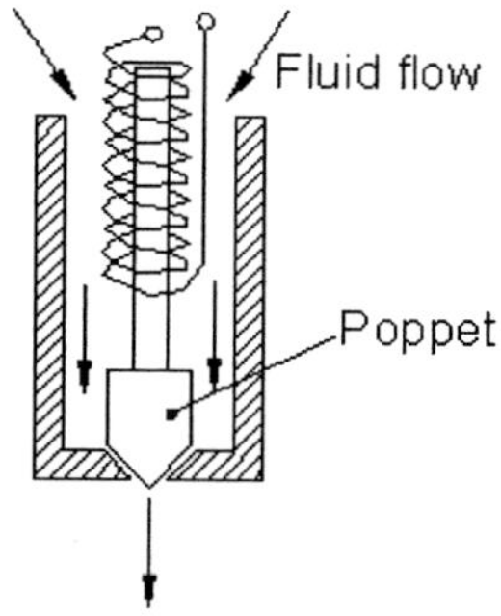

Figure 4 Poppet Valve Structure

A Poppet valve is used to open and close a hydraulic transmission line quickly. In a normally close design, the valve is closed under the action of an internal spring, which pushes the poppet against the seat, thus closing the outlet. When the solenoid of a poppet valve is energized the magnetic force from the solenoid pulls the poppet against the spring, this produces a flow path. In a normally open variation, the solenoid energizes to close the flow path.

The construction of a poppet valve is relatively simple (see Fig.4). It is easy to control, and has a short response time. The open-close operation of the valve is always in the range of a few tens of a millisecond. It is relatively free of contamination problem, as the poppet will push away the contaminants or seal over it.

3.2 Electric Actuation

Electrical motors are probably the most universal method for converting from one form of energy to another. They are best suited for payloads less than 900N. Compared with hydraulic system, their repeatability is very high. They also require less maintenance, and their noise levels are generally lower. DC servomotors, direct drive DC torque motors and stepper motors are used in robotic applications. Excellent speed regulation, high torque and high efficiency make DC servomotors the favorite choice in robotic applications. Either they allow precise control of the speed, by manipulation of the voltage, or the torque, by manipulation of current applied to the motor.

Recent commercial introduction of integrated motors has again revolutionized the way in which electric actuators are deployed. Motors integrated with gear drives, feed back sensors, and serial communication ports are making the motor application easier and design very compact. Animatics (USA) is the pioneer of integrated motion control. SmartMotor™, an integrated motor by Animatics, is a complete, compact and user-friendly integrated control system that features a brushless DC servomotor, motion controller, encoder and amplifier.

3.3 Modularity in Design

With the availability of integrated motors, the way manipulators are designed also experienced a rapid change. Re-configurable manipulators that are compact and easy to control are becoming a common occurrence. With the serial communication ports available, communication and control of joints became very easy. The hassle of rewiring and cabling of manipulators with every minor design changes is no more a hurdle in manipulator design. This has lead to the adopting of modular design concepts in robotics. Manipulator joints can be designed as modules and can be assembled as per requirement. They are more or less independent of manipulator kinematics and link dimensions. Modular design for the inspection manipulator is conceptualized using integrated motors.

4.0 Overall Conceptual Design

The functional requirements and the design constraints explained in the earlier section, warrants the design of a customised manipulator. Two of the most basic issues in the design of the manipulator are the number of degrees of freedom (d.o.f) and the resulting kinematic structure. The number of degrees-of-freedom required for a manipulator is decided on the basis of the task to be performed by the manipulator. For a general-purpose manipulator, it is quite common to have 6 d.o.f. This is to enable it to reach the desired position and orientation in space. In order to assess the requirement of number of degrees-of-freedom for the inspection manipulator, it is necessary to study the inspection task and define it as a Cartesian task, as follows.

The fundamental task of the manipulator during an inspection task is (a) to position the inspection probe at the point of inspection (b) to orient the probe at the appropriate attitude with respect to the weld seam. For the purpose of analysis, it is assumed that the weld seam is toroidal. The cross-section of such a weld seam is shown in Fig. 5. During non-contact inspection, it is necessary that the inspection tool follows the weld seam, with the tool at a proper orientation. Assume that the co-ordinate frame XYZ is attached to the inspection probe as shown in Figure 5. A local co-ordinate frame, $X_oY_oZ_o$, is attached to the inspected weld point, in such a way that the Y_o axis is normal to the weld seam. The X_o axis is tangential to the seam, while the Z_o axis completes a right-handed coordinate frame with the first two axes. For inspection, the basic requirement is that the frames of the weld seam and probe coincide. Thus, probe manipulation is a three dimensional task, which warrants a 6 d.o.f manipulator.

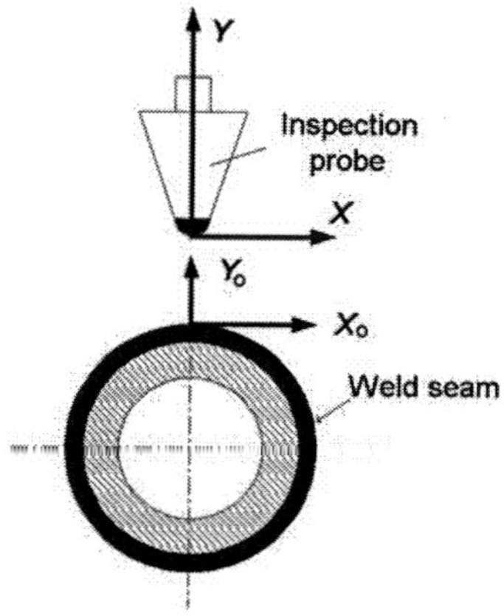

Figure 5 - Coordinate Frames Of The Tool And Weld Seam

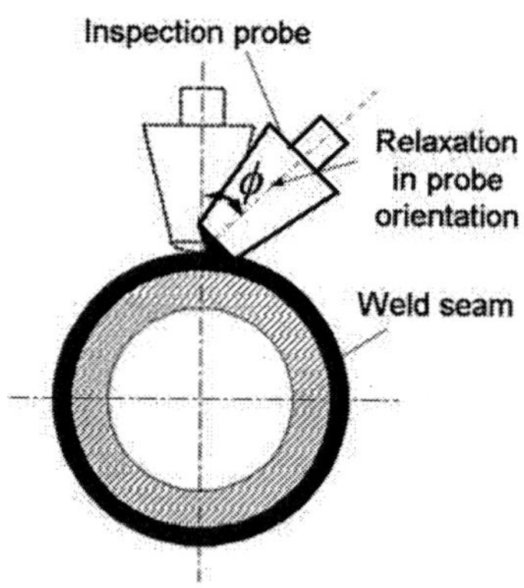

Figure 6. - Relaxation Of The Probe Orientation During Inspection

Inspection of welded structures using the ACFM procedure allows a relaxation in the orientation of the probe [8] with respect to the joint co-ordinate frame at least in one direction without affecting the quality of inspection. The experiments conducted using an Alternating Current Field Measurement (ACFM) probe for inspection of weld joints have shown that the probe orientation in one direction, say pitch, can be relaxed up to 60° as shown in Figure 6, without affecting the quality of inspection process. (The actual value of relaxation angle depends on the type of probe and the quality requirements set for the inspection task). This property of the inspection process simplifies the inspection process to a 5 d.o.f problem-3 positioning and two orientations. Thus, a manipulator with 3 d.o.f for positioning and 2 d.o.f for orientation can be used as an inspection arm for weld structures, without affecting the inspection quality.

Having established the sufficiency of a 5 d.o.f manipulator for inspection purposes, the feasibility of such a configuration for underwater manipulator was analysed. The design requirements of the manipulator specify a total length of between 2m to 2.5m and a dexterous workspace towards the boundary of the actual workspace. In order to ensure a dexterous workspace, it is necessary to have the wrist links (last two links) as small as possible and have their rotational axis intersect at a common point. If we design with short wrist links length, the lengths of the other three links will need to be large in order to ensure the required total length requirement of the manipulator. Such large link lengths will affect the manipulation adversely and result in many voids in its workspace.

Alternate options are available. One of the options at this stage is to add redundant d.o.f to the manipulator in order to reduce the link lengths, thus improving the manipulability. But, for an underwater manipulator, this is not a feasible option, as increases in the number of joints will eventually lead to increase in weight as well as additional actuators and associated systems. In order to overcome this design constraint, a novel manipulator architecture having a multi stage actuation is proposed.

4.1 Multi Stage Manipulator

A multi-stage manipulator is one having more than one distinct stage of independent actuation. This can be compared with a micro-macro manipulator concept having a large positioning manipulator carrying a small manipulator for fine manipulation. A two-stage, one launching stage and another inspection stage, configuration is proposed for the underwater manipulator. The inspection stage manipulator is responsible for carrying out all manipulation tasks like, cleaning, tracking and inspection. The launching stage will be used, primarily, to position the inspection stage close to the target workspace. Using this scheme, it is possible to increase the overall reach of the manipulator without increasing the length of the inspection stage manipulator.

As discussed in the previous section, a 5 d.o.f manipulator can adequately be used as the inspection stage manipulator, and the launching can be provided by a 2 d.o.f manipulator. Thus, a 7 d.o.f, 2 stage manipulator can be concluded to satisfy the design criteria for workspace and dexterity.

4.2 Kinematic Structure of the Manipulator

Selecting a particular configuration for a manipulator is purely a qualitative aspect of design and no rigorous reasoning is available to prove the validity of selection. The factors affecting the selection of manipulator configuration are its intended task, workspace, positional reachability, controllability of orientation, link interference, speed of actuation, folded configuration etc. The kinematic structure for the inspection stage and the launching stages are chosen based on the specific tasks they have to perform. In the case of the inspection stage manipulator, different kinematic structures were proposed and analysed for its workspace, singularities, kinematics and folded configurations. A kinematic structure as shown in Figure 4, was found to be the best option. This configuration has three joints with parallel axis of rotation and two joints with intersecting joint axis. The first three joints can position the end effector in a plane (X-Z) and orient it in that plane. The fourth joint is used for positioning the end effector in a plane orthogonal to the plane selected by the first three joints and the fifth joint is used for orienting the end effector along its axis of rotation, orthogonal to all other axes of rotation.

The kinematic structure of the first stage manipulator have been selected in such a way that it will enhance the total reachable workspace and allow an appropriate folded configuration for the combined manipulator. The first joint of the manipulator has been provided with a rotational axis parallel to the fourth joint of the second stage. This will enhance the reach of the inspection stage manipulator through the entire range of this joint. The second joint axis is parallel to the first joint of the inspection stage and this allows the positioning of the second stage in the X-Z plane. Thus, the launching stage manipulator is capable of positioning the inspection stage manipulator in a 3 dimensional space with very close proximity to the task space. The kinematic structure of the combined manipulator is shown in Figure.5. Further details on the kinematic structure and its salient features are provided [9]. Appropriate selection of joint axes for both the stages, allows a perfectly folded configuration for the combined manipulator. The folded configuration of the manipulator is shown in Figure.6. Such a configuration ensures minimum influence on the dynamics of the URV during its free flight from one location to another.

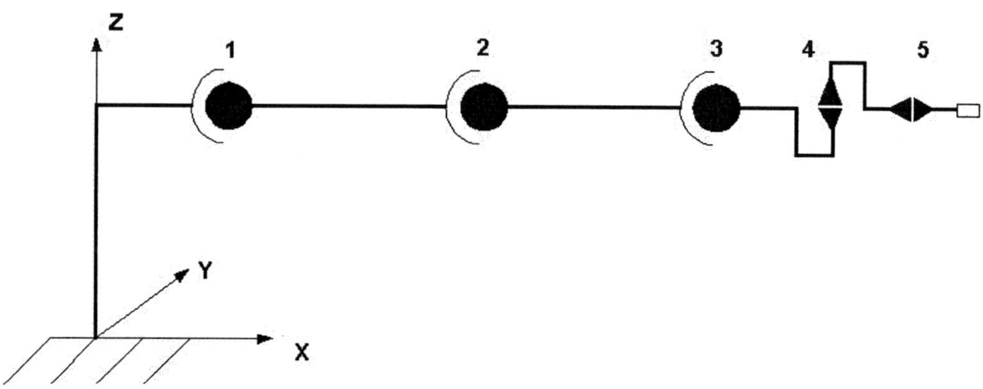

Figure 7. - Kinematic Structure Of The 5 D.O.F. Inspection Manipulator

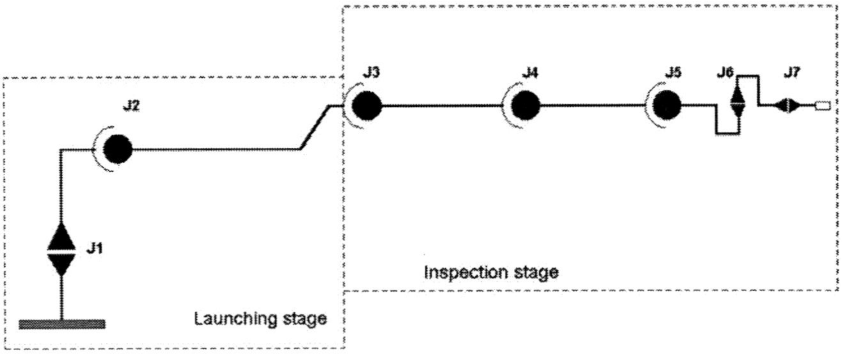

Figure 8. - Kinematic Structure Of The 7 D.O.F. Manipulator

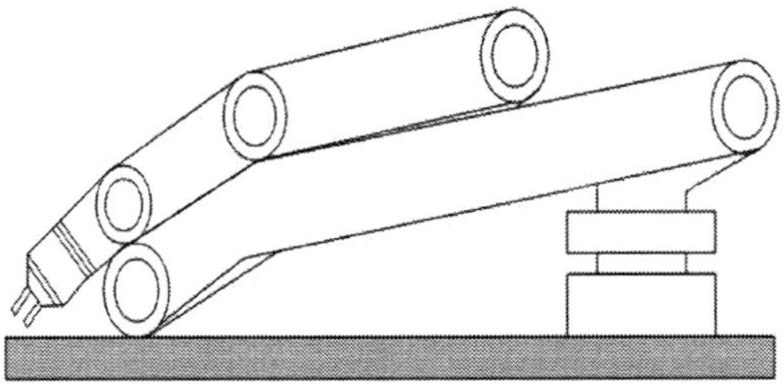

Figure 9. - Folded Configuration of the Manipulator

By introducing a launching stage, it was possible to have a large workspacé for the manipulator without increasing the link lengths for the inspection stage. Moreover, this will enable the docking of the ROV at a safe distance from the inspection site. Since the workspace required for the inspection manipulator is small compared to the total workspace, the inspection stage manipulator can have shorter link lengths, without affecting the total reach of the manipulator. This will ensure a more dexterous workspace towards the boundary of the reachable workspace of the manipulator. In addition, small link lengths will reduce the torque required to drive the links. Once the inspection manipulator is positioned with a reasonable accuracy towards its target workspace by the launching stage, the inspection manipulator will carry out the accurate positioning and inspection of the weld joint. After the inspection manipulator has completely tracked the weld joint in its operating space, the launching stage will position it to another point to resume the inspection process. The immediate benefits of this type of a configuration are:

- Very large workspace with more dexterity towards the boundary of the workspace can be achieved.
- The inspection stage manipulator can have smaller link lengths compared to the launching stage.
- The positioning accuracy required for the launching stage is relatively low compared to the inspection stage.
- The launching stage is to be operated only for a short duration at frequent intervals of time during an inspection process, while the inspection stage has to be operated continuously.

- Because of the large link lengths and the load due to the inspection manipulator, the torque required to drive the launching stage is very high.
- The launching stage requires high power actuators, which are to be operated for a short duration, while the inspection stage requires relatively low torque actuators, which are to be operated for longer duration.

4.3 Hybrid Power Structure

As explained in section 3, the accuracy and manipulation capabilities of the manipulator depend on the type of actuators used for actuating the joints. In order to reap the benefits of both the hydraulic as well as electric actuators, a hybrid power structure has been adopted for the subsea manipulator. Since the inspection stage manipulator requires high positional accuracy, electrical actuators have been selected for this stage. Integrated DC servomotors with RS485C serial communication protocol have been selected. A modular design for the inspection stage is proposed. These drives are expected to give good positional accuracy, as the load on the actuators is relatively low due to the short link lengths and payload. High torque hydraulic actuators have been selected to drive the first stage manipulator. The use of high torque hydraulic actuators for this stage matches the operating needs of large torque and low positioning accuracy requirements of the launching stage. Such a hybrid power structure is expected to be highly effective in the deployment of the inspection stage manipulator.

5.0 Kinematic Analysis

The geometry of arm motion and the spatial configuration of the manipulator, as a function of time, are analysed. The direct and inverse kinematic problems, workspace analysis, and link length optimisation are presented.

5.1.1 Direct Kinematics

Direct kinematics of the manipulator can be found by adopting the classical Denavit-Hartenberg convention. Co-ordinate frames, as shown in Figure.10, are assigned to the manipulator joints and the D-H parameters of the manipulator are determined as given in Table 1.

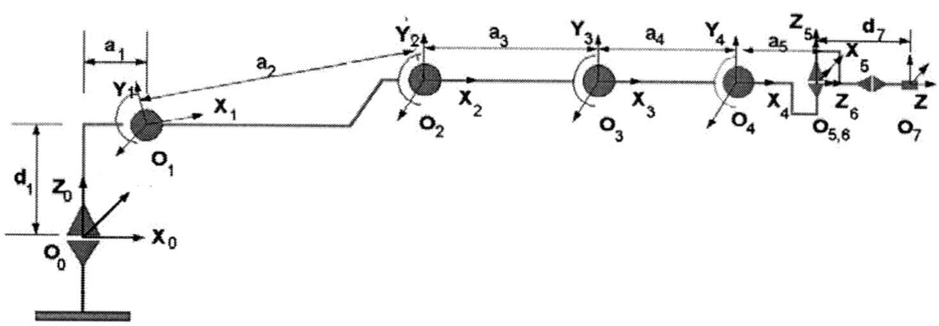

Figure 10. - Assignment Of Coordinate Frames

Homogeneous transformation matrices expressing relative position and orientation between consecutive links are computed and the resulting direct kinematics equation can be expressed in the form

$$T(q) = \begin{pmatrix} n(q) & s(q) & a(q) & p(q) \\ 0 & 0 & 0 & 1 \end{pmatrix} \tag{1}$$

where, q is the (7x1) vector of joint variables, and n, s, a are the unit vectors of the end-effector frame and p is the position vector of the origin of such frame with respect to the origin of the base frame.

5.1.2 Inverse Kinematics

The inverse kinematics of the manipulator is solved in two stages. In the first stage, the joint angle θ_1 and θ_2 for the launching stage is solved using the position vector p(q) and orientation vector a(q). The second stage solution consists of solving for the joint angles θ_3 to θ_7 for the inspection stage manipulator with θ_1 and θ_2 as known parameters. Since the second stage manipulator has three adjacent joint axes parallel, a closed form solution can be easily found. The closed form solution procedure for the manipulator can be found in [10].

5.1.3 Joint Angle Limits and Workspace

The operational limits of the joints have been fixed on the basis of the functional requirements as well as mechanical constraints. The joint limits for all the joints are given in Table 2. The operational workspace of the manipulator can be considered as a union of the workspaces of the launching stage and the inspection stage. The workspace of the inspection stage will be circular in cross section in the X-Z plane. The total workspace can be obtained by rotating this subspace through the limits of joint angle 2 in the X-Z plane and that of joint angle 1 in the Y-Z plane, as shown in Figure 11.

Table 1. - Link Lengths and Operational limits of joints

Link	Joint variable, θ	α	Joint limits, °	a 'mm'	d 'mm'	Length, mm
1	θ1	90	-90 to + 90	75	150	-
2	θ2	0	-90 to +135	700	0	700
3	θ3	0	-120 to +135	502	0	502
4	θ4	0	-120 to +135	355	0	355
5	θ5	-90	-120 to +135	192	0	192
6	θ6	90	-60 to + 60	0	0	150
7	θ7	0	-180 to +180	0	300	125

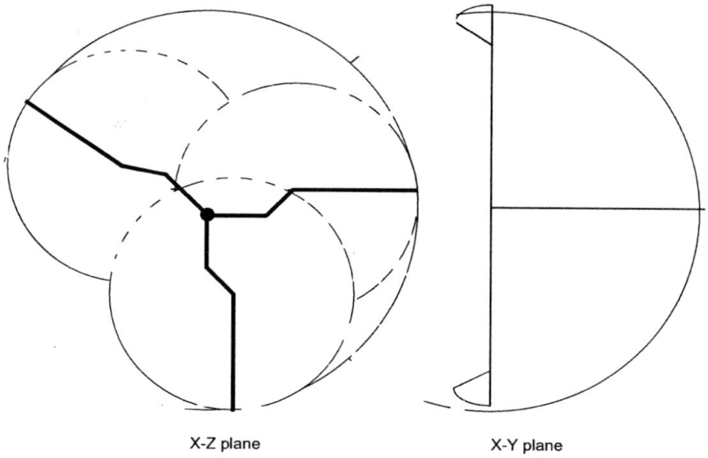

Figure 11. - Manipulator Workspace

5.1.4 Link Length Optimization

A few methods [11,12] have been reported in the literature for the optimization of manipulator link lengths, most of them dealing with a particular objective of optimum workspace, dexterity, etc., but these cannot be readily adapted to our case. Dexterity of the manipulator at a given target point is heavily dependent on the kinematic structure and link lengths of the manipulator. In this paper, the optimum link lengths of the manipulator are found on the basis of highest dexterity for any given maximum length of the manipulator. From the architecture of the inspection stage manipulator, it is clear that the first three d.o.f. of this manipulator are used for positioning and orienting the end-effector in the X-Z plane, as depicted in Fig. 7. The fourth joint allows the end-effector to go out of this plane, while the last joint allows the orientation of the end-effector in a plane normal to the X-Z plane. Therefore, the task of the first three links of the inspection manipulator reduces to the positioning of the sixth joint center O_6 in the X-Z plane. The lengths of the last two links are chosen on the basis of maximum allowed relaxation in the tool orientation and the mechanical constraints of the joints. Therefore, these links are designed based on considerations beyond dexterity. The remaining links of the inspection manipulator are optimized for maximum dexterity.

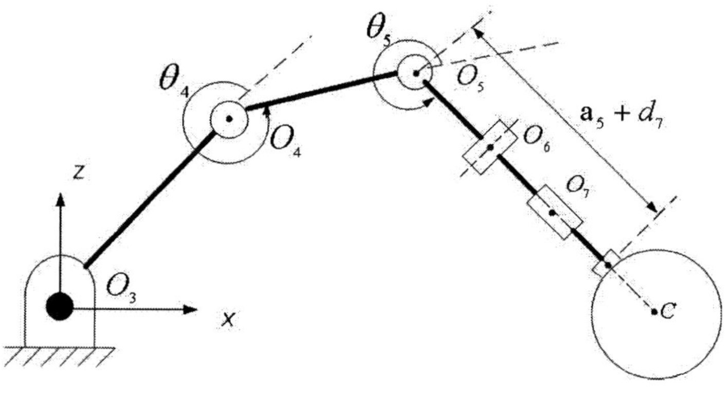

Figure 12. - Inspection Of A Welded Seam Using 5 D.O.F. Manipulator

As a measure of manipulator dexterity, we adopt here the condition number [13] of the inspection manipulator, based on the condition number of the Jacobian matrix. The condition number of a square matrix is a measure of the relative round-off error amplification of the computed results, upon solving a linear system of equations, associated with that matrix, with respect to the relative round-off error of the data [14]. Likewise, the condition number of a rectangular matrix measures the round-off error amplification of the data upon solving a rectangular system of linear equations using a generalized inverse [14]. The condition number is known to vary between unity and infinity, matrices with a unity condition number being called isotropic. For a manipulator Jacobian to attain isotropy, the architecture of the manipulator must be designed accordingly. A manipulator so designed is said to have kinematic or, more generally, kinetostatic isotropy, such manipulators being termed isotropic. Kinematic isotropy was used by Salisbury and Craig [15] for the optimum design of mechanical fingers. An extension of isotropy to other manipulators is given in [16]. Kinematic isotropy has been used widely in the design of serial and parallel manipulators [17, 18].

Consider the inspection manipulator in contact with the outer circle of a weld-seam surface for inspection, as shown in Fig.13. It has been shown by Salisburry and Craig [15] that for a 2 d.o.f manipulator to be isotropic, the ratio between its link lengths must be

$$\frac{l_1}{l_2} = \frac{\sqrt{2}}{2} \tag{2}$$

Assuming that the first two link lengths are in the ratio $a_4/a_3 = \sqrt{2}/2$, the relationship among joint angles is derived below.

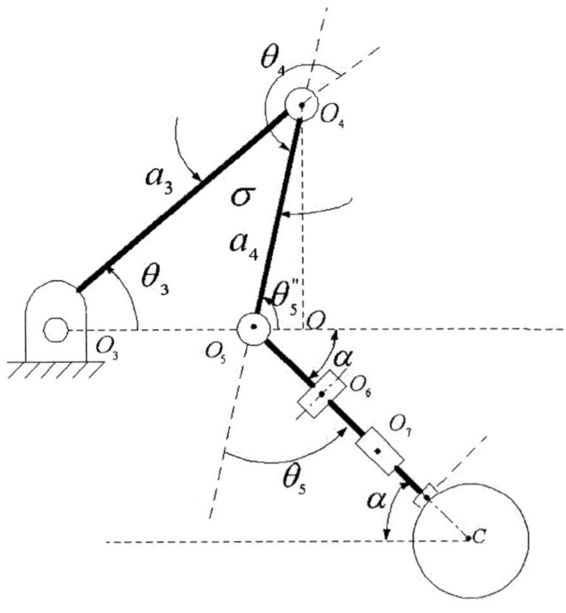

Figure 13. - Isotropic Configuration Of The Manipulator

$$\sigma = \theta_4 - \pi \tag{3}$$

$$\theta_5 = \pi - (\theta_5^{''} + \alpha) \tag{4}$$

Like wise, for triangle $O_3O_4O_5$

$$\theta_3 = \theta_5^{''} - \sigma \tag{5}$$

From the right angled triangle O_3O_4O, moreover

$$a_3 \sin\theta_3 = a_4 \sin\theta_5^{''} \tag{6}$$

Substituting θ_3 form equation (5) into equation (6) gives

$$\frac{\sin\left(\theta_5^{''} - \sigma\right)}{\sin\theta_5^{''}} = \frac{a_4}{a_3} = \frac{1}{\sqrt{2}} \tag{7}$$

or

$$\sin\theta_5^{''} = \sqrt{2}\sin\left(\theta_5^{''} - \sigma\right) \tag{8}$$

Solving for $\theta_5^{''}$ from equation (8) gives two possible values for $\theta_5^{''}$ lying π radians apart, namely

$$\theta_5^{''} = \arctan\left(\frac{x}{y}\right) \tag{9}$$

where,

$$x = \sqrt{2}\sin\sigma, \quad y = \sqrt{2}\cos\sigma - 1 \tag{10}$$

Thus, for any known value of angle σ, within the working region, all the other joint angles can be calculated using equations (3), (4), (5) and (9) for a known contact angle α. This angle depends, in turn, on the location of the weld seam with respect to the manipulator, and is highly uncertain. This uncertainty in contact angle can be addressed by considering the range of variation of the contact angle and solving the equations for the worst-case scenario of inspection. A limiting case of the manipulator posture for inspection, as shown in Figure 14, is considered. In this posture, the second link will lay tangent to the weld seam and the contact angle α will be 0° to ensure stable manipulator contact with the weld seam. The minimum link length requirements to achieve this posture can be determined by the method given below.

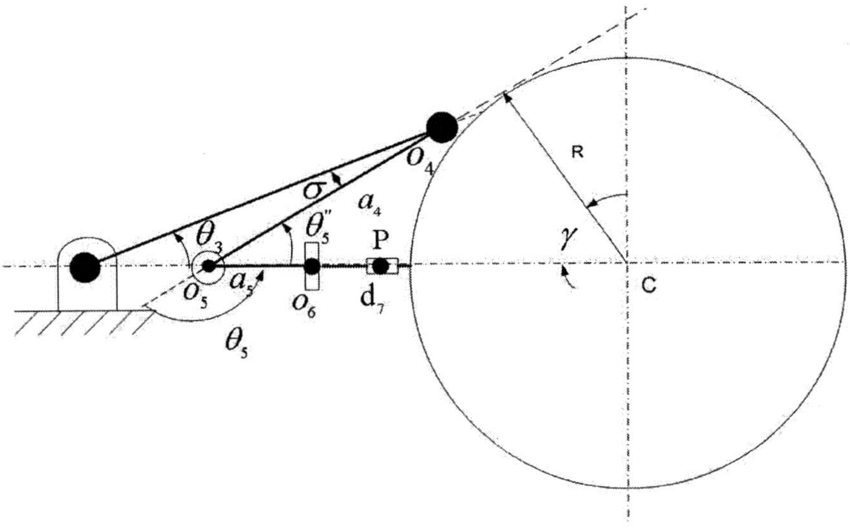

Figure 14. - Worst-Case Posture Of The Manipulator During Inspection

The radius of the largest weld seam to be inspected may be assumed to be R. Angle σ is assumed, moreover, to be at its lower limit for this posture of the manipulator. By virtue of the tangency assumed, angle γ can be calculated as

$$\gamma = \frac{\pi}{2} - \theta_5'' \tag{11}$$

The minimum link lengths required to achieve the posture at hand are found below. From the right triangle with hypotenuse $\overline{O_5C}$

$$\overline{O_5C} = \frac{R}{\cos\gamma} \tag{12}$$

$$\overline{O_5P} = \frac{R}{\cos\gamma} - R \tag{13}$$

$$\overline{O_5O_6} = a_5 = \overline{O_5P} - \overline{O_5O_6} = \overline{O_5P} - d_7 \tag{14}$$

Using eqs. (16) and (17), it is possible to estimate the minimum link length a_5 required to establish a stable contact with the weld seam for the worst case scenario of inspection. Any value of $\sigma > 10^{o}$ and/or $\alpha > 0^{o}$ will ensure a better condition number for the manipulator Jacobian and hence better dexterity for the manipulator at the inspection point than that at the worst case posture.

It is observed from the above relations that the diameter of the weld seam plays an important role in determining the link lengths. Since the weld diameter is a variable, which cannot be predicted in advance, it is necessary to fix the maximum size of the seam to be inspected by the manipulator. The minimum link lengths estimated using this value would always ensure a better condition number for any size of the weld seam smaller than the maximum, by positioning the links away from the singular configurations, and hence better dexterity at the inspection point. By fixing the maximum length of the manipulator as 2050mm and fixing the maximum weld seam diameter as 1000mm, link lengths for the manipulator was estimated using the above procedure. The estimated link lengths are shown in Table 1. (For more details on link length optimisation, please refer [19]).

6.0 Torque Analysis

In order to size the actuators for various joints, it is necessary to analyse the torque requirements at every joint. Static torque due to gravitational and external load and dynamic torque due to acceleration needs to be considered. Dynamic equation for manipulator has the general form

$$\tau = M(\Theta)\ddot{\Theta} + V\left(\Theta, \dot{\Theta}\right) + G(\Theta) \tag{15}$$

where, τ is the n x 1 vector giving the total torque acting, M(Θ) is the n x n mass matrix of the manipulator, V(Θ,Θ) is an n x 1 vector of Coriolis and centrifugal terms, and G(Θ) is an n x 1 vector of gravity and external load terms. Since the underwater manipulator is a slow moving system, torque due to acceleration and Coriolis effect is negligible when compared to the gravitational and external load torque. Hence they are neglected in the preliminary design analysis. Gravitational and external load torque can be calculated, using the classical methods available, as follows:

$$G(\Theta) = \tau_g + \tau_l \tag{16}$$

where, τ_g is the torque due to gravitational load and τ_l is the torque due to external loading .

Gravitational torque acting at joint 'i' of an '*n*' link manipulator can be calculated as:

$$\tau_{g(i)} = -\sum_{j=i}^{n} m_j g^T \, {}^i\mathbf{J}_j \tag{17}$$

where,

- m_j = Mass (including link mass, actuator mass etc.) of jth link acting through its CG.
- g = [g_x g_y g_z] = gravity vector
- ${}^iJ_j$ = Jacobian of jth link CG w.r.t. i^{th} link frame; $j >= i$.

For an External load, $F_e = [\ Fx\ Fy\ Fz\ ]^T$, acting at the tip of the manipulator, the corresponding joint torques acting on the joints are given by

$$\tau_l = J^T \left[{}^n R_1\right] F_e \tag{18}$$

where, R is the rotational matrix representing the nth joint frame with respect to the base frame. From the assumed mass of links and actuators, and an estimated external load, the initial estimate of torques at various joints of the manipulator shown in fig. 6, was calculated using the above equations. The estimated torques, for the posture shown in Fig.6, is shown in Table 2.

Table 2 - Estimated Joint torques

Joint	Link and Actuator mass (N)	Estimated Joint Torque, Nm. (Gravitational and External Load)
1	50	0
2	80	350
3	70	200
4	60	50
5	40	10
6	15	0
7	10	0

As can be seen from the table, joints 2 to 5 experience gravity and external loads. The actuators for these joints are to be chosen on the basis of the joint torques. The remaining actuators are chosen based on acceleration torque experienced at the joints. For an assumed acceleration of 0.1 rad/s^2, it is found that the torque requirement at joint 1 is approximately 50 Nm.

7.0 Link Design

Underwater manipulator links have to accommodate numerous and frequently, conflicting constraints. Some of these constraints are

- The link should have an internal hollow area to provide conduits for electric power and communication cables, hoses, power-transmitting components, control rods etc. Also, these cavities can be used for pressure balancing during deep-sea manipulation.
- At the same time, their external dimensions are limited by the overall design constraints.
- Links have to be as light as possible to reduce inertia forces and allow for the largest payload per given size of motors and actuators.
- For a given weight, links should have highest possible bending stiffness.

The structural dynamic characteristics of manipulators include stiffness, inertia, damping, and natural frequencies. Stresses in the links of manipulating robots are relatively low. Stiffness is the most important general criteria of manipulator link design. The problem of stiffness enhancement is especially important since conventional techniques, such as "beefing up" cross-section or using high modulus conventional material, are in many cases not acceptable because, mostly, they are counterproductive. Larger cross-section brings with it weight increases, higher inertia forces and thus lager deflection. The most effective ways of achieving reduction of inertia are through judicious selection of material and/or cross-section shape as well as an optimized combination.

Good load carrying capacity, less deflection under static loads, good hydrodynamic characteristics, and better corrosion resistance are basic requirements of the manipulator link. Link geometry plays a vital role in achieving some of these performance criteria. Three cross-sections, square, rectangular, and circular, were considered for the links and analysed for its load bearing and deflection characteristics.

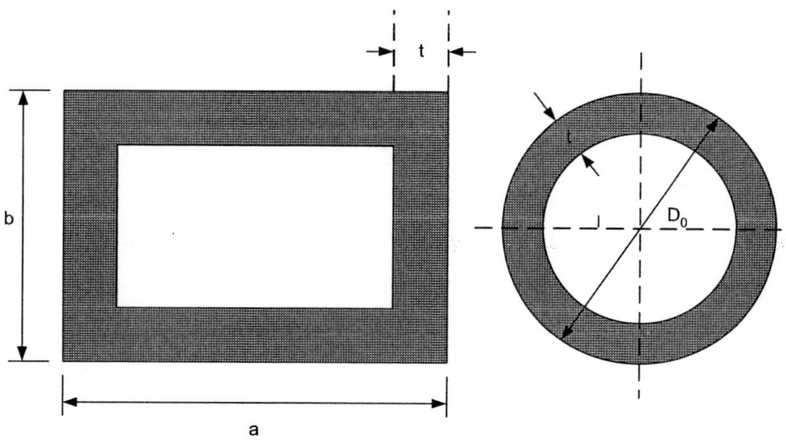

Figure 15 - Cross Section Of Rectangular And Cylindrical Link

For the hollow circular cross section, the moment of inertia, $I_{rd,}$ and cross sectional area, $A_{rd,}$ are given as:

$$I_{rd} = \frac{\pi(D_0^4 - D_1^4)}{64} = \frac{\pi\left[D_0^4 - (D_0 - 2t)^4\right]}{64}$$
$$\cong \frac{\pi D_0^3 t}{8}\left(1 - 3\frac{t}{D_0} + 4\frac{t^2}{D_0^2}\right) \qquad (19)$$
$$A_{rd} = \frac{\pi\left(D_0^2 - D_1^2\right)}{4} = \pi D_0 t\left(1 - \frac{t}{D_0}\right)$$

For the rectangular cross-section, the value of I depends on the direction of the neutral axis in relation to which the moment of inertia is computed.

$$I_{rec,x} = \frac{\left(ab^3 - (a-2t)(b-2t)^3\right)}{12}$$
$$I_{rec,y} = \frac{\left(a^3b - (a-2t)^3(b-2t)\right)}{12} \tag{20}$$

For the square cross-section,

$$I_{sq} = \frac{a^4}{12} - \frac{(a-2t)^4}{12} \cong \frac{2}{3}a^3t\left(1 - 3\frac{t}{a} + 4\frac{t^2}{a^2}\right) \tag{21}$$

The cross sectional area of the rectangular and square cross sections, respectively, is:

$$A_{re} = ab - (a-2t)(b-2t) = 2t(a+b) - 4t^2$$
$$A_{sq} = 4at(1 - \frac{t}{a}) \tag{22}$$

If we assume the outer dimensions and wall thickness to be same for all cross sections, for an assumed wall thickness of $t=0.2D_o$, we have:

$D_0 = a$;

$$A_{rd} = 0.503 \; D_0^2 \qquad I_{rd} = 0.0405 \; D_0^4$$
$$A_{sq} = 0.64 \; a^2 \qquad I_{sq} = 0.0705 \; a^4$$
$$\frac{A_{sq}}{A_{rd}} = 1.27 \qquad \frac{I_{sq}}{I_{rd}} = 1.84$$

If $t = 0.1 \, D_0$, then:

$$A_{rd} = 0.283 \; D_0^2 \qquad I_{rd} = 0.029 \; D_0^4$$
$$A_{sq} = 0.36 \; a^2 \qquad I_{sq} = 0.0493 \; a^4$$
$$\frac{A_{sq}}{A_{rd}} = 1.27 \qquad \frac{I_{sq}}{I_{rd}} = 1.70$$

Thus, a square cross section provide a 69 percent to 84 present increase in rigidity with only a 27 percent increase in weight; its effectiveness is higher for higher wall thickness. Thus, for the same weight, a beam with thin-walled square cross sections would have 40 to 60 percent higher stiffness.

Since the links for the underwater manipulator are loaded in one direction, a rectangular cross section with higher moment of inertia along the loading axis was chosen. A series of finite element modeling and analysis of various possible combinations of size and materials were conducted to choose the best possible link size and material. Weight of the link, deflection under load, and suitability of application were the criteria for the final selection. The final choice of link dimensions and materials is shown in Table 3.

Table 3 - Link Cross-Section And Dimensions

Link	**Material**	**Cross section**	**Size (mm)**	**Max. deflection under estimated load (mm)**
1	Stainless Steel	Circular	OD=40, ID=30	0.0
2	Titanium	Rectangular	30 x 60, t=3.2	0.8mm
3	Titanium	Rectangular	30 x 60, t=3.2	0.5mm
4	Titanium	Rectangular	30 x 50, t=3.2	0.15
5	Titanium	Rectangular	30 x 40, t=3.2	0.05
6,7	Titanium	Circular	OD=30, ID=25	0.015

8.0 Actuators and Control

Two hydraulic actuators and five electric motors were selected as drives for the manipulator. The torque requirements for the actuator are given in Table 2. Since linear hydraulic actuators are not suitable for the folded configuration of the manipulator, rotary vane type hydraulic actuators were chosen for the hydraulic stage. The hydraulic actuators chosen are:

Joint 1: HKS DAH 40-180°, with additional radial bearings, Max. Pressure 210 bar.
Joint 2: DS-Dynatec HPSD 110-280°; Max. Pressure 140 bar.

Poppet type, on-off valves, was chosen as the flow control valves. Four valves (HydraForce, SL08-22) were used for each joint and they were actuated using a micro controller for Pulse Width Modulated on-off control. PWM control strategy is discussed in section 5. Integrated DC servomotors are chosen for the electrical stage. Animatics Corporation's SMARTMOTOR™ is selected as drives for 3rd to 6th Joint of the manipulator. They are integrated motors with feedback sensors and serial communication ports. This helps to reduce the cabling requirements between the manipulator controller and individual motors. The 7th joint is powered using a custom designed integrated motor. The controller structure is designed on the basis of the actuators chosen and the proposed switching strategy. The designed architecture for the manipulator controller is shown in Fig. 16.

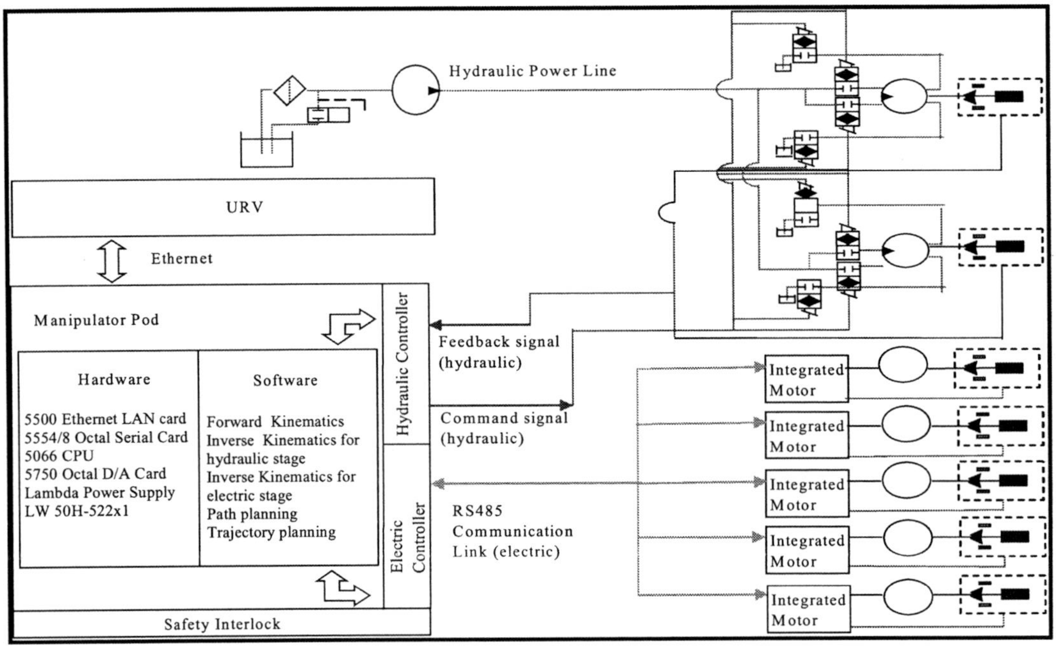

Fig. 16 - Controller Architecture For Manipulator

9.0 Manipulator Design

9.1 Hydraulic Stage

Mechanical design of the hydraulic circuit and hydraulic joints were carried out. The hydraulic circuit for the manipulator joint is shown in Fig. 17. An accumulator is used in the circuit to meet the peak demand flow rates. Design drawings for the hydraulic joints are shown in Fig. 18.

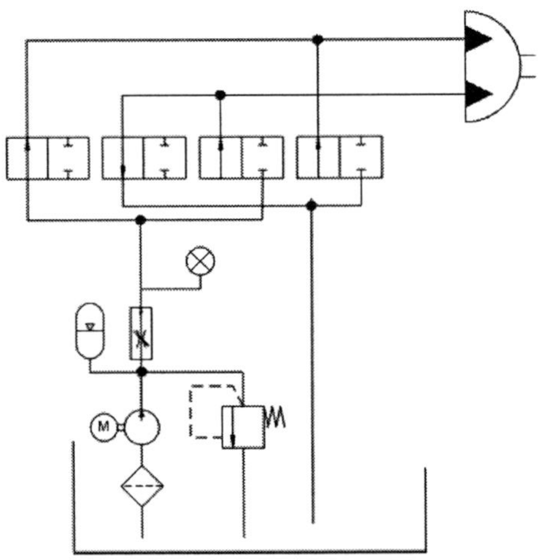

Fig. 17 - Simplified Hydraulic Circuit For One Actuator

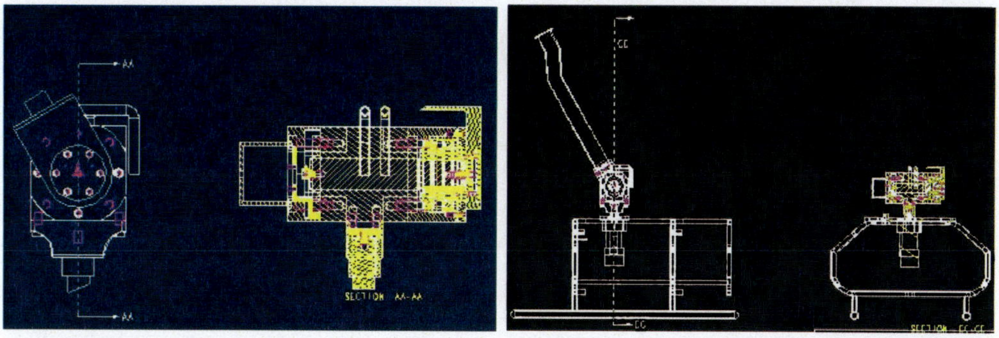

(a) Assembly Drawing Of First And Second Joints (b) Assembly Drawing Of Launching Stage
Fig 18 (a) - Manipulator On URV frame

9.2 Electrical Stage

Design of electrical stage manipulator was carried out in accordance with the design requirements established earlier. Images of "ProEngineer" CAD model of the 5 d.o.f manipulator and design drawing for one of the joints are shown in Figure 19.

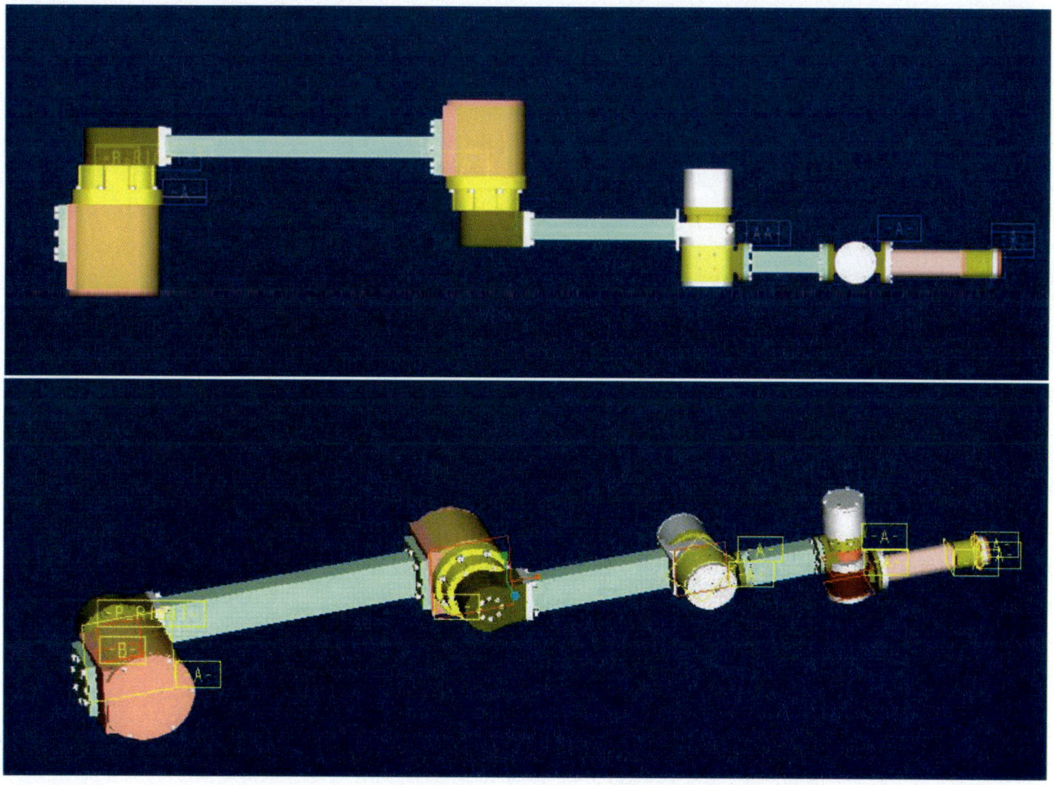

Fig. 19a - CAD Model Of Inspection Stage Manipulator

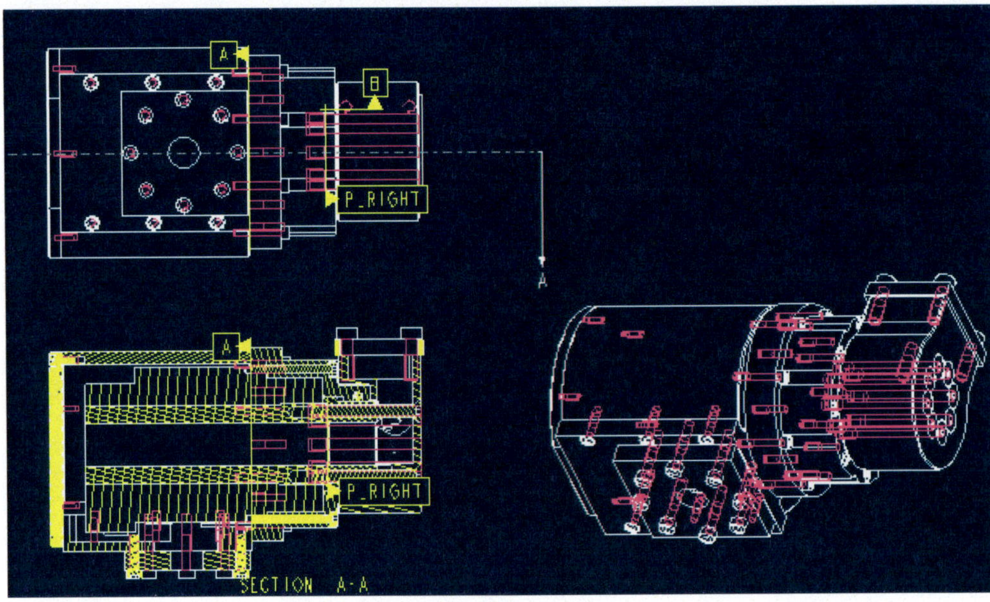

Fig. 19b - Design Drawing An Electrical Joint

10.0 Switching of Poppet Valve

The standard method of controlling hydraulic actuators using a poppet valve is to energise the poppet valve for flow when motion is required, and de-energise the valve when the actuator has reached its desired position. More sophisticated switching methods have been attempted in an effort to provide a smoother and controlled acceleration and retardation motion. These switching schemes have been made available through the use of low cost microprocessors. PWM switching of the energizing current to the poppet solenoid is another attempt at enhancing the valve operation by lowering the heating of the solenoid and bring about savings in electrical power. Solenoid operated poppet valves are widely used in on/off control of hydraulic systems. They are easy to control, simple in construction, and less prone to contamination. They are suited for rugged applications where the positional accuracy is not critical.

10.1 Modeling and Simulation

In order to establish a switching strategy for the actuators, modeling and simulation of the poppet valves and actuators were carried out. Simulink, the simulation tool box from MATLAB, was used for modeling and simulation. The model of the poppet valve is shown in Fig. 18(a). The simulation response of the valve for a 500 millisecond period pulse actuation is shown in Fig. 18 (b). It was found that the response time of the valve is approximately 20 milliseconds to open and 30 milliseconds to close. The flow output depends on the pressure differential existing at the poppet side. Poppet valve parameters are chosen from the manufacturer's catalogue (Hydra Force, SL08-22).

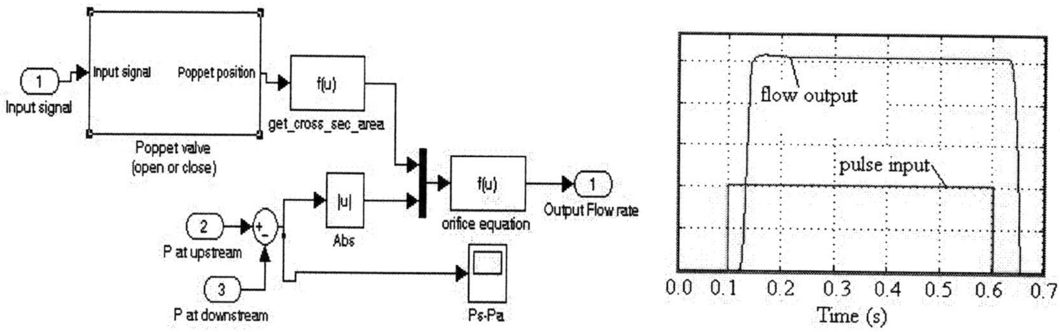

(a) (b)

Fig. 18. a) - Poppet Valve 'Simulink' Model Fig. b) - Simulation Result For A Step Input

The Simulink model of one of the hydraulic joints is shown in Fig.19. Four poppet valves are used for controlling each actuator. The response of the actuator for an input pulse of 500m.sec is shown in Fig. 20. It was found that the response time of the actuator is around 30 milliseconds to start moving and 50 milliseconds to stop. The response of the actuator for a continuous pulse train of 100 Hz at 50% duty cycle is shown in Fig. 21.

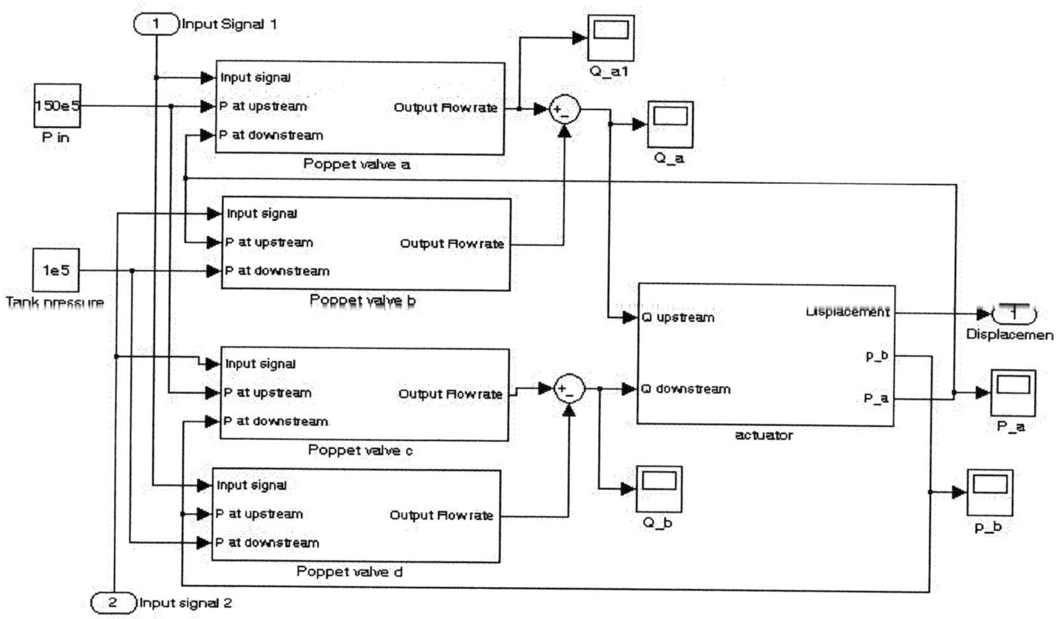

Fig. 19 - Simulink Model of The Hydraulic Joint

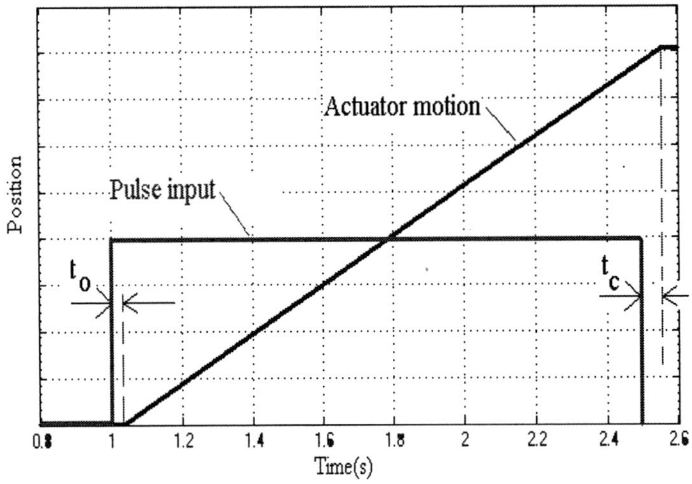

Fig. 20 - Response Of The Joint For A Pulse Input

10.2 Experimental Verification

Experiments were carried out to study the actuator response to various input commands. Fig. 21 shows the response of the actuator for a pulse input. The response time of the actuator is around 50 milliseconds to start and around 70 milliseconds to stop. This is slightly more than that obtained in simulation. The increase in response time may be attributed to differences between the actual and applied parameter values for bulk modulus, viscous friction etc. The result, however, shows the predicted nature of the actuator response and correlates well.

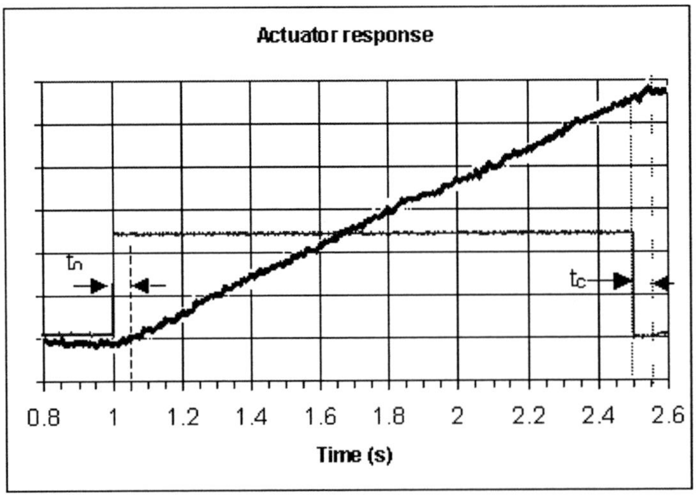

Fig. 21 - Actuator Response For A Single Pulse Input

11.0 Conclusion

Design challenges of an underwater inspection manipulator were discussed in this paper. Special design requirements of a customized inspection manipulator were analysed and a hybridly powered multistage configuration was proposed. Kinematic design analysis, optimisation of link lengths and controller architecture have been presented. Modeling and simulation of a hydraulic joint for Pulse Width Modulation (PWM) control using poppet valves has been presented. Simulation results have established the validity of the simple model. This has been verified witha experimental studies. Using the basic model, more sophisticated switching techniques may be studied to enhance controllability. Research works in this area are progressing with the objective of enhancing positioning accuracy and providing smoother joint motions.

References

[1] LeTessier, R.; Coade, T.R.; and Geneve, B. *Int. Nat. J. Pressure vessels and piping.* 2002, vol. 79, 549-554.

[2] Barry, Fallon; Stevan, B. S.; Charles, F.R.; William, S.G. URSULA: Design of an underwater manipulator for Nuclear Reactor vessel Inspection, Robotics for challenging Environments, *Proceedings of ASCE Specialty Conference*, New Mexico, 1994,pp 311-319.

[3] Greig, A.R. and Broome, D.R. Development of a Robotic underwater Manipulator, *Transactions of Int. nat .Marine Engineers*, Vol. 106, Part 5, 1994, pp 217-229.

[4] Andrew J Taylor: Design issues for underwater manipulator systems, *Mechatronics*, Vol.3, No.4, 1993, pp 419-432.

[5] Reddy, D.V.; and Arockiasamy, M. [Eds.]: Offshore Structures , Volume 1, Krieger Publishing, Florida 1991,

[6] Francis A Giuliano [Ed.]: *Introduction to Oil and Gas Technology*, Prentice Hall, NewJersy.1989.

[7] Esposito Anthony, *Fluid Power with Applications*, Prentice Hall, 2003.

[8] Low C.K. *Defect evaluation using ACFM technique*, Project report, Nanyang Technological University, Singapore, 2000.

[9] Asokan, T.; Seet, G., Iastrebov, V.; and Senanayake, R. Kinematic design and analysis of a 7 degree of freedom dual stage inspection manipulator for dexterous sub sea applications, *J. of Intelligent and Robotic Systems*, **38**:277-295, 2003.

[10] Asokan, T.; Seet, G. and Choudhury, S. Kinematic Analysis of a 7 d.o.f hybrid powered multi stage manipulator for under water applications, *Proc. of Int. Nat. Conference on Computational Intelligence, Robotics and Autonomous systems*, Nov. 28-30, Singapore, 106-110, 2001.

[11] Ceccarelli, M. Optimal design and location of manipulators. In *Computational Dynamics in Multibody Systems* (Eds. M.F.O. Seabra Pereira and J.A.C. Ambrosio), Kluwer Academic, Boston, 131-146, 1995.

[12] Vijaykimar, R.; Waldron, K.J. and Tsai, M.J. Geometric optimisation of serial chain manipulator structures for working volume and dexterity, *Int. J. Rob. Res.*, **5**(2), 91-103, 1986.

[13] Angeles, *J. Fundamentals of Robotic systems*, Theory, Methods and Algorithms, Springer Verlag, New York, 1997.

[14] Golub, G.H. and Van Loan, C.F. *Matrix Computations*, The Johns Hopkins University press, Baltimore, 1989.

[15] Salisburry, J.K. and Craig, J.J. Articulated hands: force control and kinematic issues, *Int. J. Rob. Res.*, **1**(1), 4-7, 1982.

[16] Angeles, J. and Lopez-Cajun, C.S. Kinematic Isotropy and the conditioning index of serial manipulators, *Int. J. Rob. Res.*, **11**(6), 560-571, 1992.

[17] Lee, J.; Duffy, J. and Hunt, K. H. A practical quality index based on the octahedral manipulator, *Int. J. Rob. Res.*, **17**(10), 1081-1090, 1998.

[18] Ranjabaran, F.; Angeles, J. and Gonzalez-Palacios, M. A. The mathematical design of a seven-axis manipulator with kinematic isotropy, *J. Intelligent and Robotic Systems*, **14**(1), 21-41,1995.

[19] Asokan, T.; Seet, G. and Angeles, J. The optimum dimensioning of an underwater manipulator for weld inspection, *Proc. IMechE., Part M: J. Engineering for the maritime Environment,* Vol. 217, 177-184, 2003.

In: Frontiers in Robotics Research
Editor: Max A. Denket, pp. 67-111
ISBN 1-60021-097-X

Chapter 3

ON THE SELECTION OF BAND-PASS QUADRATURE FILTERS*

Djamal Boukerroui†
EUDIASYC, UMR CNRS 6599, Université de Technologie de Compiègne, BP 20529 - 60205 Compiègne Cedex, France
J. Alison Noble‡ ***and Michael Brady***§
Department of Engineering Science,
University of Oxford, Oxford OX1 3PJ, UK.

Abstract

Band-pass quadrature filters are extensively used in computer vision to estimate information from images such as: phase, energy, frequency and orientation, possibly at different scales and utilise this in further processing-tasks. Arguably, local[1] phase is the most desirable information since it encodes most of the relevant perceptual information in an image. Phase-based image analysis techniques have been used for estimating optical flow, matching, registration and feature detection. The estimation is intrinsically noisy and depends critically on the choice of the quadrature filters. In this chapter, we first study the mathematical properties of the quadrature filter pairs most commonly seen in the literature and then consider some new pairs derived from the classical feature detection literature. In the case of feature detection, we present the first attempt to design a quadrature pair based on filters derived for optimal edge/line detection. A comparison of the filters is presented in terms of feature detection performance, wherever possible, in the sense of Canny and in terms of phase stability. We conclude with remarks on how our analysis can aid in the choice of a filter pair for a given image processing task.

---

*Parts of this work appeared previously in [4]. The authors thank Kluwer academic publishers and Springer/the language of science for permission to reproduce the material which appeared previously in that publication.

†E-mail address: djamal.boukerroui@hds.utc.fr

‡E-mail address: noble@robots.ox.ac.uk

§E-mail address: jmb@robots.ox.ac.uk

[1]Throughout, we mean "local" or "instantaneous"... When 1-D signals are discussed, the "space" variable may be called "time".

1 Introduction

In computer vision and image analysis, arguably the most classical processing tools are amplitude-based techniques, both in the spatial and frequency domains. Consequently, throughout the history of digital image processing, smoothing and differentiation have been subjects of intense study. A variety of optimal differential operators have been proposed to solve different computer vision problems (see for example: [9, 16, 17, 45, 62, 75, 81, 82, 94, 96] for edge and line detection and [2, 19, 84] for optical flow estimation). Differentiation is highly sensitive to noise, but can be reduced/avoided using an appropriate scale selection for the smoothing function. However, amplitude-based measurements are sensitive to smooth shading and lighting variations. In particular, this can be a source of serious error, for example for differential optical flow estimation. Moreover, an image contains many types of edges (not only step-edges). For instance, for edge and line detection, one might, for example, combine the outputs of a step detector and line detector. However, this is not a perfect solution since it leads to edge duplication (see for e.g. [79, 96]). Furthermore, especially for medical images such as X-ray mammograms [39], the interesting features in the image often do not have the idealised edge shape; here, we do not only mean blur degradation but also feature combination (for example feature number 4 in Fig.1 is a combination of a line with a positive step edge). These may considerably reduce the optimality of the filter and lead to unpredictable multiple responses.

Phase-based feature detection has been investigated extensively following the publication of the Local Energy (LE) model of feature detection [67]. This model postulates that features can be defined and classified using their phase signatures (or their phase congruency (PC)). This observation have led to the development of a number of phase-based feature detection algorithms [23, 36, 46, 68, 72, 73, 76–79, 89]. Phase-based feature detection (PC & LE) provides a *single* unified theory for detecting a wide range of features, rather than being specialised for a single feature type such as intensity steps.

Phase information has also been used successfully for matching and velocity estimation. For instance, phase-difference [44] and phase-correlation [53] techniques have been used for discrete two-view matching and phase gradient used for the measurement of image orientation and optical flow [28, 68] and recently in image registration [38, 64].

Numerous additional advantages of the use of phase-based measurements have been reported. Perhaps the most desirable property of phase is that it is theoretically invariant to brightness and contrast. Hence it is, in principle, robust against typical variations in image formation. However, the estimation of phase is intrinsically noisy and depends critically on the choice of the quadrature filters. It is this estimation that we investigate. The key contribution of this chapter is to define quantitative performance metrics for quadrature filters to enable a meaningful comparison of filter options to be made and hence to guide the choice of filters in practice.

1.1 Local Phase and Amplitude in 1D Signals

Mathematically, the local properties (Amplitude and Phase) of a 1D real signal (function) $f(x)$ are defined using the so called analytical signal (function) $f_A(x)$ [32]:

$$f_A(x) = f(x) - i f_{\mathcal{H}}(x); \quad (1)$$

where $i = \sqrt{-1}$ and $f_{\mathcal{H}}(x)$ is the Hilbert transform of $f(x)$ defined by :

$$f_{\mathcal{H}}(x) = \frac{1}{\pi}\int_{-\infty}^{\infty}\frac{f(\tau)}{\tau - x}\,d\tau \;\Leftrightarrow\; F_{\mathcal{H}}(\omega) = F(\omega)\cdot i\,\mathrm{sign}(\omega)\ , \tag{2}$$

where, $F(\omega)$ is the Fourier transform[2] of $f(x)$ and

$$\mathrm{sign}(\omega) = \begin{cases} -1 & \omega < 0 \\ +1 & \omega \geq 0 \end{cases}$$

It is straightforward from (1) and (2) to see that:

$$F_A(\omega) = F(\omega)\cdot[1 + \mathrm{sign}(\omega)]\ . \tag{3}$$

Hence, the analytical signal corresponding to f is obtained by suppressing all its negative frequencies and multiplying all its positive frequencies by two. Given the definition of the analytical signal, we can make the following definitions:
the local amplitude[3] of $f(x)$:

$$A(x) = \|f_A(x)\| = \sqrt{f^2(x) + f_{\mathcal{H}}^2(x)}\ . \tag{4}$$

and the local phase by:

$$\phi(x) = \arg(f_A(x)) = \arctan\{f(x)/f_{\mathcal{H}}(x)\}\ . \tag{5}$$

From the above definition of local phase we can derive a number of phase signatures for different image features (see for e.g. [32, 89]). However, direct calculation of these local quantities, as defined above, cannot be done in a phase-based technique. This is mainly because the Hilbert transform/analytical signal, as one can see from equation (2) and (3), is defined over the whole signal/spectrum of the signal; while, for example, for feature detection, localisation in both space and frequency are highly desirable. In other words, we have to design an operator to approximate these quantities in a small, spatial span and over a narrow range of frequencies (scales) to enhance spatial localisation and to avoid the effect of noise. One way is to 'window' the function f, $f_w(x) = w(x) \times f(x)$, using, for example, a Gaussian function as $w(x)$ leading to the Short-Time Fourier Transform (STFT) (see for e.g. [60, chap. 4]).
An alternative approach is to estimate the local phase/energy of a filtered version of the signal. One property which this should satisfy is a zero response for a constant signal (to be invariant to grey level shift [78]), which implies that the approximation filter $f_e(x)$ must be:

1. a band-pass filter (zero-DC).

[2] Throughout, we use the following convention: if $f(x)$ is a 1D function then $F(\omega)$ is its Fourier transform.

[3] Normally, energy is defined as the amplitude-squared [31, 78], however, some authors (see for e.g. [49, 66, 77]), define the energy as being the amplitude.

2. symmetric (with constant phase, so as not to change the phase information of the original signal). The symmetry condition implies that the filter must be even (hence which is why it is denoted f_e).

We then have:

$$\begin{aligned} \hat{f}_A(x) &= f_e(x) \circledast f(x) - i\,\mathcal{H}(f_e(x) \circledast f(x)) \\ &= (f_e(x) - i\,\mathcal{H}(f_e(x)) \circledast f(x) \\ &= (f_e(x) - i\,f_o(x)) \circledast f(x) \ . \end{aligned} \tag{6}$$

where "$\circledast$" denotes the 1-D convolution operator and $f_o(x)$ is the Hilbert transform of $f_e(x)$, hence they are in quadrature. It can be shown, given the definition of the Hilbert transform and the Hermitian property of the Fourier transform, that $f_o(x)$ is an odd, zero-DC band-pass filter (see e.g. [32]).

In practice, the computation of an approximation of the local phase and the local amplitude uses a pair of band-pass quadrature filters (an odd filter $f_o(x)$ and an even filter $f_e(x)$), where:

$$\hat{\phi}(x) = \arctan\{f_e(x) \circledast f(x) / f_o(x) \circledast f(x)\} \ , \tag{7}$$

$$\hat{A}(x) = \sqrt{[f_e(x) \circledast f(x)]^2 + [f_o(x) \circledast f(x)]^2} \ . \tag{8}$$

An example of an amplitude scalogram and a phase scalogram of a 1D signal is shown in Fig. 1. The signal is a combination of different type of edges; the first part of the signal is an ideal profile and the last part is a noisy version of the signal (additive white Gaussian noise with a signal-to-noise = 18 dB). Observe that all the features, independently of their shapes, are detected and can be classified using their phases. However the interpretation of a type of feature is obviously scale dependent. Consider for example the second feature: inspecting the amplitude scalogram we notice that at small scales, the feature is detected as a positive step edge followed by a negative step edge; however at very large scales there is only one peak in the amplitude scalogram and the corresponding phase at this maximum is the phase of a line. Notice for example the first feature is no longer detected at large scales; the amplitude scalogram shows a black spot and the corresponding phase is not defined, this point in scale space is called a singularity point.

1.2 Multidimensional Signals

The analytical signal and the corresponding quadrature filters provides a powerful framework for the extraction of local properties of signals. Until recently, two (and more generally N-) dimensional signal theory suffered from the absence of an *isotropic* extension of the analytical signal/Hilbert transform. Several attempts to generalise the Hilbert transform may be found in the literature [24]: for instance, the directional/partial Hilbert transform [32, 79], the total Hilbert transform [34], and a combination of these two in the Fourier domain [34] and in the quaternionic Fourier domain [7]. Unfortunately, none of these approaches satisfy the isotropy property, which is necessary to obtain invariance with respect to orientation. Hence, other implementations have been proposed in the literature, for two-dimensional signals via the use of steerable filters [30, 85] or by performing the analysis at a

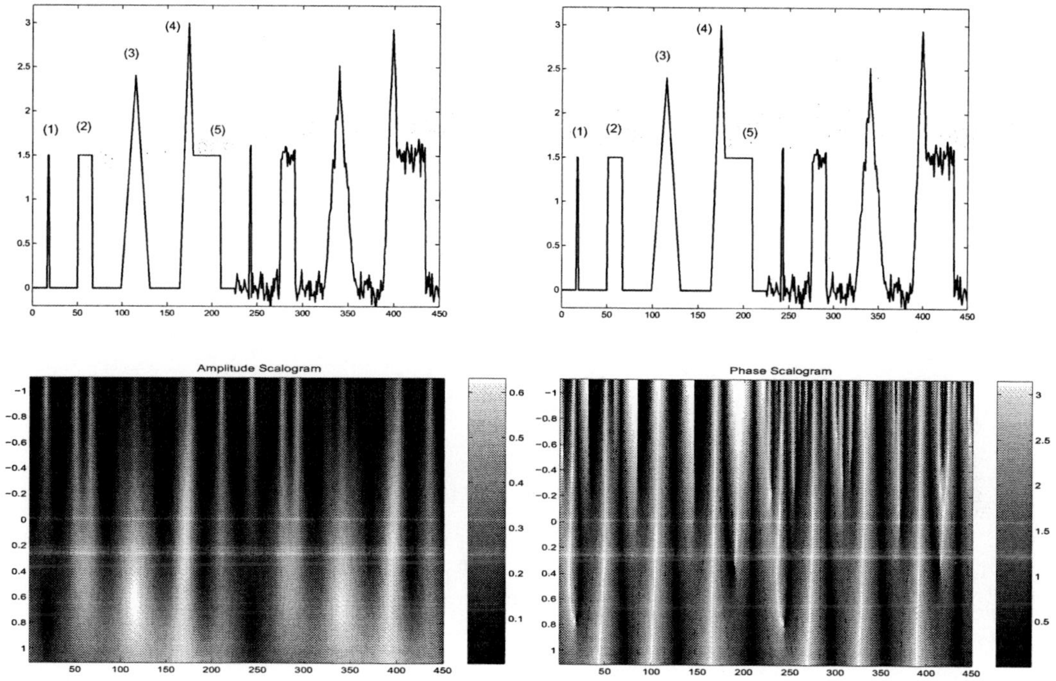

Figure 1: Example of a 1D signal and scalograms of local amplitude (left) and the absolute value of local phase (right). The level contours correspond to a local phase of $\mp\frac{\pi}{2}$ radians. A local phase of $-\frac{\pi}{2}$, $+\frac{\pi}{2}$, 0 and π corresponds to a valley, a roof(line/bar), a negative step edge and a positive step edge respectively. The scalograms are computed using a difference of Gaussian filter of 2 octaves bandwidth. The vertical axis shows the wavelength of the filter in a log scale $ln(\frac{\lambda_0}{50})$. The top row is in duplicate deliberately to help see the correspondences.

set of orientations, then combining the results to provide the information at any orientation in the image [35, 36, 49, 56, 66, 68, 76, 77, 89].

Recently, Felsberg and Sommer [22, 24] proposed a novel two-dimensional generalisation of the analytic signal based on the Riesz transform, which is used instead of the Hilbert transform. A 2D isotropic analytic signal, the *monogenic signal*, is proposed. This representation preserves the core properties of the 1D analytic signal that decomposes a signal into information about structure (local phase) and energy (local amplitude). The basic idea is to design an odd isotropic filter which is vector-valued rather than scalar-valued. The "monogenic signal" is an important "phasor" for describing image properties (local phase, local amplitude and local orientation). The construction of the monogenic signal is straightforward given the 1D quadrature filter; the reader is referred to [24] for more detail.

Several applications of the monogenic signal have been already realised: edge detection and local orientation estimation [23], Moiré interferograms [54], image denoising in a scale space framework [25] and image disparity estimation [21]. A unifying approach to phase based image processing in scale space, the "monogenic scale-space", is proposed in [26].

1.3 Summary

Having introduced the analytical signal, and its extension to the multidimensional case (the monogenic signal), we have a powerful mathematical framework for the estimation of local properties of signals. These properties could be used in a number of different image processing applications. In practice, the estimation framework uses a quadrature pair of band-pass zero-DC linear filters. The construction of the multidimensional pair is straightforward given the 1D pair, or in fact given just one of the filters (the odd or the even). The question which naturally arises is: how should one choose the quadrature pair for a given application. This is the question that is addressed in the remainder of this chapter.

2 Common Quadrature Filters

In this section, we study a representative set of the most common quadrature filters in the literature[4] as well as number of new pairs derived from classical feature detectors. We provide the reader with either a mathematical expression of filter properties or a numerical estimation. For simplicity, occasionally, for a given property but different filters, we use the same symbol.

2.1 Gabor Filters

In 1946, Denis Gabor [31] introduced the windowed Fourier transform, also called the Short Time Fourier Transform (STFT). Essentially, instead of using a sinusoidal basis function $e^{i\omega x}$ which is well localised in frequency but not localised in time, he used what are now called "Gabor atoms". These are constructed by translating in time/space and frequency a real and symmetric time/space window g:

$$g_{x_0,\omega_0}(x) = g(x - x_0)\exp[i\omega_0 x] \ . \tag{9}$$

The energy of g_{x_0,ω_0} is concentrated in the neighbourhood of x_0 over an interval which can be measured by the standard deviation σ_g of $||g||^2$. Its Fourier transform is a translation by ω_0 of the Fourier transform G of g. Therefore the energy of g_{x_0,ω_0} is concentrated near ω_0, over an interval of size σ_G. In a time-frequency (scale-space) plane (x, ω), the energy spread of Gabor atoms can be represented as a rectangle of width (σ_g, σ_G) centred on (x_0, ω_0). Such a rectangle is called a "Heisenberg rectangle"[60] and the uncertainty principle shows that its area satisfies:

$$\sigma_g \sigma_G \geq \frac{1}{2} \ . \tag{10}$$

Equality applies only when g is a Gaussian, in which case the atoms g_{x_0,ω_0} are called Gabor functions.

Gabor functions are used extensively in digital image processing because of their optimality in the sense of the uncertainty principle, and because of their similarity to the

[4]Of course, this is not an exhaustive list of quadrature filters in the literature. We did not include, for example, *shiftable* [83] filters as this property implies their use in a multiscale framework, the robust quadrature filters proposed in [63] as well as some recent filters [26, 48].

responses of simple cells in the primary visual cortex of primates [13, 61, 69, 70]. They have been used, for instance, in texture analysis [1, 5, 18, 43, 51], texture segmentation [33, 41, 43, 90, 91], feature detection [10, 11, 66, 74], motion estimation [28, 44], image representation and coding [55], and segmentation from motion [92]. Recently, Young et al. [93] proposed a fast recursive implementation of Gabor filters.
A one-dimensional Gabor function is a Gaussian modulated complex exponential:

$$g_{b\,(\sigma,\,\omega_0)}(x) = n_c \frac{1}{\sqrt{2\pi}\,\sigma} \exp\left(-\frac{x^2}{2\,\sigma^2}\right) \exp[i\,\omega_0\,x] \ , \tag{11}$$

$$G_{b\,(\sigma,\,\omega_0)}(\omega) = n_c \exp\left(-\frac{\sigma^2}{2}(\omega-\omega_0)^2\right) \ , \tag{12}$$

where $\omega_0 = 2\pi/\lambda_0$ is the peak tuning frequency and n_c is a normalisation constant which is application dependent. Two alternative normalisation conditions have been used most frequently in the literature[5]:

1. maximum condition:

$$\max_{\omega} G_b(\omega) = 1 \quad \Rightarrow n_c = 1 \ . \tag{13}$$

2. constant energy condition:

$$\|g_b(x)\| = \int_{-\infty}^{\infty} g_b^{\star}(x)\, g_b(x)\, dx = 1 \quad \Rightarrow n_c = \sqrt{2\,\sqrt{\pi}\,\sigma} \ . \tag{14}$$

The half-response spatial frequency bandwidth β (in octaves) is defined as:

$$\beta = \log_2\left(\frac{\omega_2}{\omega_1}\right) \ , \tag{15}$$

where ω_1 and ω_2 are the solutions of $G_b(\omega)|_{n_c=1} = 1/2$ and $\omega_1 < \omega_2$. It follows that the bandwidth β of the Gabor filter(s) is the following function of $k_\beta = \sigma\,\omega_0$:

$$\beta(k_\beta) = \frac{\ln\left(\frac{k_\beta+\sqrt{2\,\ln(2)}}{k_\beta-\sqrt{2\,\ln(2)}}\right)}{\ln(2)} \ ; \tag{16}$$

alternatively,

$$k_\beta = \sqrt{2\,\ln(2)}\,\frac{2^\beta+1}{2^\beta-1} \ . \tag{17}$$

A complex scalar $z(\Delta x,\, \Delta\lambda)$ encodes the phase behaviour of the Gabor kernel, in the presence of white Gaussian noise, for a small spatial perturbation $\Delta\,x = x_1 - x_0$, and a small scale perturbation $\Delta\,\lambda = \lambda_1 - \lambda_0$, in the neighbourhood of a scale-space point $\mathbf{p}_0(x_0 = 0, \lambda_0)$:

$$z_{g_b}(\Delta x,\, \Delta\lambda) = \sqrt{\frac{2\lambda_1\,\lambda_0}{\lambda_1^2+\lambda_0^2}} \exp\left(-\frac{k_\beta^2(\lambda_1-\lambda_0)^2}{2(\lambda_1^2+\lambda_0^2)} - \frac{2\pi^2\,\Delta x^2}{k_\beta^2\,(\lambda_1^2+\lambda_0^2)} + i\,2\pi\,\frac{\lambda_1+\lambda_0}{\lambda_1^2+\lambda_0^2}\,\Delta x\right) \ . \tag{18}$$

[5]Note: the use of these two normalisation conditions is a recurrent theme throughout.

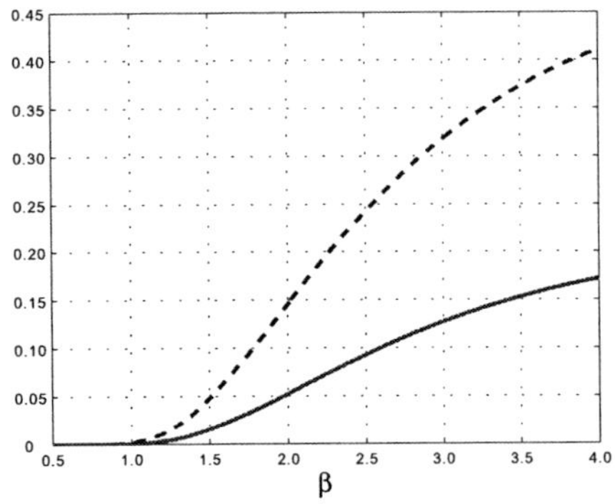

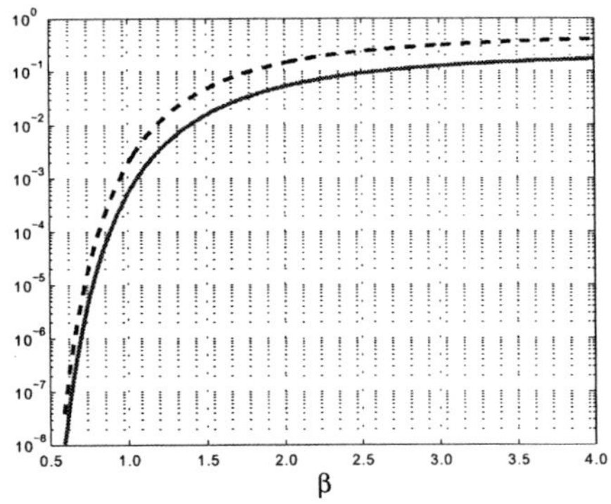

Figure 2: Evolution of DC_{g_b} (dashed curve) and the negative energy $\sqrt{\mathrm{NE}_{g_b}}$ (solid curve) of Gabor filters as a function of bandwidth. Linear scale (left); logarithmic scale (right).

See also Appendix A.

Unfortunately, Gabor functions have several limitations in terms of their direct application in computer vision. First, Gabor filters are non-orthogonal [32] though the family forms a frame (the family is complete but redundant)[55]. Although Gabor functions are well localised, they inevitably have infinite support. Truncation is used in most practical implementations to avoid aliasing. Finally, a Gabor pair is not in quadrature as it has non-zero negative frequencies, and a non-zero DC component (the cosine component does not integrate to zero): The normalised negative energy is:

$$\mathrm{NE}_{g_b} = \frac{\int_{-\infty}^{0} G_b(\omega)^2\, d\omega}{\int_{-\infty}^{\infty} G_b(\omega)^2\, d\omega} = \frac{1}{2}(1 - \mathrm{erf}(\sigma\,\omega_0)) = \frac{1}{2}(1 - \mathrm{erf}(k_\beta)) \ , \tag{19}$$

and the DC component is:

$$\mathrm{DC}_{g_b} = G_b(0)|_{n_c=1} = \exp\left(-\frac{\sigma^2\,\omega_0^2}{2}\right) = \exp\left(-\frac{k_\beta^2}{2}\right) \ . \tag{20}$$

Note that both the negative power and the DC component are functions of the bandwidth β. Figure 2 shows the evolution of DC_{g_b} and the root of the negative energy $\sqrt{\mathrm{NE}_{g_b}}$ as functions of the bandwidth. We note that for a small bandwidth ($\beta < 0.7$) DC correction is not needed, as its value is close to the precision of computers (10^{-6}). However, for filters with a broad bandwidth ($\beta > 0.7$), a correction is needed. In the case of feature detection, Heitger et al. [37] and Ronse [78] independently pointed out this problem. Ronse [78] gave a second condition for the filter to be invariant to the addition of a linear ramp. Both authors proposed a DC correction for the even Gabor filter. We studied and compared both corrections. This is summarised in Appendix C where we also proposed a new variant to Ronse's correction.

2.2 Log-Gabor Filters

Log-Gabor (log-Normal) filters have become the preferred choice in many computer vision applications using quadrature filters as they allow arbitrarily large bandwidth zero-DC filters to be constructed [22, 23, 27, 32, 49, 68], These filters are particularly suited for the estimation of the local frequency using the ratio of log-Gabor filters [32].

The one-dimensional log-Gabor filter in the frequency domain is a Gaussian function on a logarithmic scale:

$$G_l(\omega) = n_c \exp\left(-\frac{\ln^2(\omega/\omega_0)}{2\ln^2(\kappa_\beta)}\right) , \tag{21}$$

where ω_0 is the peak tuning frequency and $0 < \kappa_\beta < 1$ is related to the bandwidth β of the filter by the following relation:

$$\beta = -\frac{2\sqrt{2}}{\sqrt{\ln 2}} \ln(\kappa_\beta) , \tag{22}$$

and inversely:

$$\kappa_\beta = \exp\left(-\frac{1}{4}\sqrt{2\ln(2)}\,\beta\right) . \tag{23}$$

The two values for the normalisation constant n_c (defined by eqs. (13) and (14)) are given by:

$$\max_\omega G_l(\omega) = 1 \quad \Rightarrow n_c = 1 , \tag{24}$$

$$\|g_l(x)\| = \frac{1}{2\pi}\|G_l(\omega)\| = 1 \quad \Rightarrow n_c = \exp\left(-\frac{1}{8}\ln(\kappa_\beta)^2\right)\sqrt{-\frac{2\sqrt{\pi}}{\omega_0\ln(\kappa_\beta)}} . \tag{25}$$

The main advantage of log-Gabor filters is that as well as being invariant to a grey-level shift (zero-DC) they are also invariant to an additive ramp [6]. A linear operator $f(x)$ is invariant to an additive ramp if and only if:

$$\int_{-\infty}^{\infty} x\, f(x)\, dx = 0 . \tag{26}$$

Using the moment property of the Fourier transform, we have:

$$\int_{-\infty}^{\infty} x\, g_l(x)\, dx = i\,\frac{d}{d\omega}\, G_l(\omega)|_{\omega=0} = 0 . \tag{27}$$

Care must be taken in the design of log-Gabor filters to avoid aliasing. Indeed, the long tails of the filters may run beyond the Nyquist limit imposed by the sampling frequency. Hence, we would like the Nyquist limit to be at least n times the standard deviation away from ω_0 on a logarithmic scale. This gives the following condition:

$$\omega_0 < \pi\,\kappa_\beta^n . \tag{28}$$

In practice, $n = 3$ is usually sufficient [32].

In addition to the above limit for the central frequency, probably the most significant drawback is the absence of an analytical expression for the filter in the spatial domain.

[6] Strictly speaking, log-Gabor filters are invariant to an additive polynomial function of order $n < \infty$ (i.e it has an infinite number of vanishing moments).

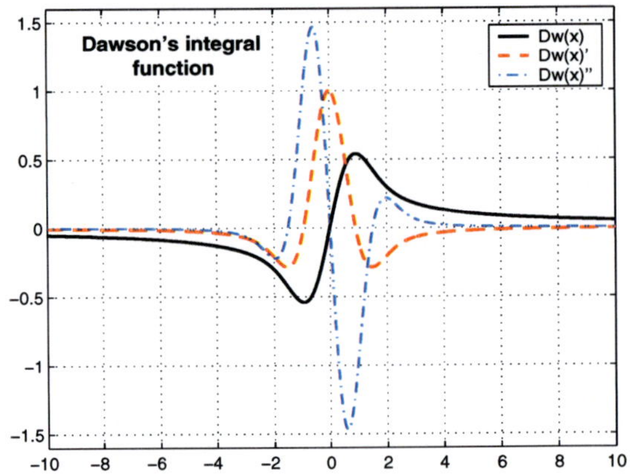

Figure 3: The *Dawson's integral* function and its derivatives in the spatial domain.

2.3 Gaussian Derivative Filters

Gaussian derivatives are well known in computer vision. This is because of some very desirable properties of the Gaussian function (used as the smoothing function), such as simplicity and steerability [30] and fast recursive implementation [15, 88]. For instance, Gaussian derivatives have been used in feature detection [9, 20, 30, 45, 56, 76, 78, 79] and motion estimation [40, 58].

It is well known that the Fourier transform of a Gaussian function is a Gaussian, and using the derivative property of the Fourier transform, we define the following family of Gaussian derivative quadrature filters in the frequency domain:

$$G_d(\omega) = \begin{cases} n_c\,\omega^a \exp[-\sigma^2\omega^2]\,, & \text{if } \omega \geq 0\,; \\ 0\,, & \text{otherwise}\,. \end{cases} \quad \text{where,} \quad a \in \mathbb{R} \quad \text{and} \quad a \geq 1\ . \tag{29}$$

Here, we relax the constraint on a and hence we do not have an analytical expression of the filters in the temporal/spatial domain for $a \notin \mathbb{N}$.

The Hilbert transform of the Gaussian function is the following function (e.g. see [78, 79]):

$$g_{\mathcal{H}}(x) = \frac{\sqrt{2}}{\pi\,\sigma}\,\mathrm{Dw}\left(\frac{x}{\sqrt{2}\,\sigma}\right)\,, \tag{30}$$

$$\text{where} \quad \mathrm{Dw}(x) = \exp(-x^2)\int_0^x \exp(t^2)\,dt \quad \text{is the \textit{Dawson's integral} function} \tag{31}$$

$$\text{and} \quad \frac{d}{dx}\mathrm{Dw}(x) = 1 - 2\,x\,\mathrm{Dw}(x)\ . \tag{32}$$

The *Dawson's integral* function and its derivatives are shown in Fig. 3. When $a \in \mathbb{N} \subset \mathbb{R}$ and odd, the corresponding odd/even filter in the spatial domain is the ath derivative of the Gaussian/$g_{\mathcal{H}}(x)$, and for even values of a the corresponding odd/even filter is the ath derivative of $g_{\mathcal{H}}(x)$/Gaussian function. This is up to a change of sign and a normalisation factor of $n_c/2$.

It is straightforward to see that the family has zero-DC, and that it is therefore invariant to a grey level shift, and for $a > 1$, it is invariant to an additive ramp (see (26)). The two values for the normalisation constant n_c are given by:

$$\max_{\omega} G_d(\omega) = 1 \quad \Rightarrow n_c = 1/G_d(\omega_0) \ , \tag{33}$$

$$\|g_d(x)\| = \frac{1}{2\pi}\|G_d(\omega)\| = 1 \quad \Rightarrow n_c = 2\,\frac{\sqrt{\pi}\,\sigma^{(a+\frac{1}{2})}}{\sqrt{\Gamma(a+\frac{1}{2})}} \ , \tag{34}$$

where $\Gamma(\cdot)$ is the gamma function and ω_0 is the peak tuning frequency given by:

$$\omega_0 = \frac{\sqrt{a}}{\sigma} \ . \tag{35}$$

The octave bandwidth of this family is given by :

$$\beta = \frac{\ln\left(\frac{\mathrm{W}(-1,\,c)}{\mathrm{W}(0,\,c)}\right)}{2\,\ln(2)} \ , \qquad \text{with} \quad c = -\frac{4^{-(1/a)}}{e} \ . \tag{36}$$

Here, $\mathrm{W}(k,\,\cdot)$ is the kth branch of the Lambert's W function [12].

The Gaussian derivative kernels, as defined here, have the advantage of allowing a continuous bandwidth with a maximum bandwidth of 2.59 octaves corresponding to the 1st derivative. Note that the analytical expression in the spatial domain is only valid for integer derivative values. Differential Gaussian filters exhibit less aliasing than log-Gabor filters.

2.4 Difference of Gaussian Filter

The Difference of Gaussian (DoG) filter has been used in a number of feature detection applications [50, 57, 59, 62, 65] and has been proposed as a filtering model of the human visual system [52, 62]. For example, retinal ganglion cells and lateral geniculate cells early in the visual processing have receptive fields which can be implemented as DoG filters [62, and references therein].

The filter (even) is given by:

$$f_{DoGe}(x) = \frac{n_c}{\sqrt{2\pi}\,\sigma'}\exp\left(-\frac{x^2}{2\,\sigma'^2}\right) - \frac{n_c}{\sqrt{2\pi}\,\sigma}\exp\left(-\frac{x^2}{2\,\sigma^2}\right) \ , \tag{37}$$

where $\sigma > \sigma' > 0$.

This operator approximates well the second derivative of a Gaussian and has one more degree of freedom[7]. Several authors have specified a preference for it, possibly because of its ease of implementation. However, this filter has a far better performance, for line detection, than the 2nd derivative of a Gaussian in the sense of Canny's criteria.

As already noted, the Hilbert transform of the Gaussian function is the Dawson's function given by (30). Hence the odd filter in quadrature with $f_{DoGe}(x)$ is given by the following:

$$f_{DoGo}(x) = n_c\frac{\sqrt{2}}{\pi\,\sigma'}\,\mathrm{Dw}\left(\frac{x}{\sqrt{2}\,\sigma'}\right) - n_c\frac{\sqrt{2}}{\pi\,\sigma}\,\mathrm{Dw}\left(\frac{x}{\sqrt{2}\,\sigma}\right) \ . \tag{38}$$

[7]Mathematically, the 2nd derivative of a Gaussian is well approximated by the DoG filter when $\sigma' \to \sigma$. However, the DoG filter preferred by Marr and Hildreth [62] is for $\sigma = 1.6\,\sigma'$.

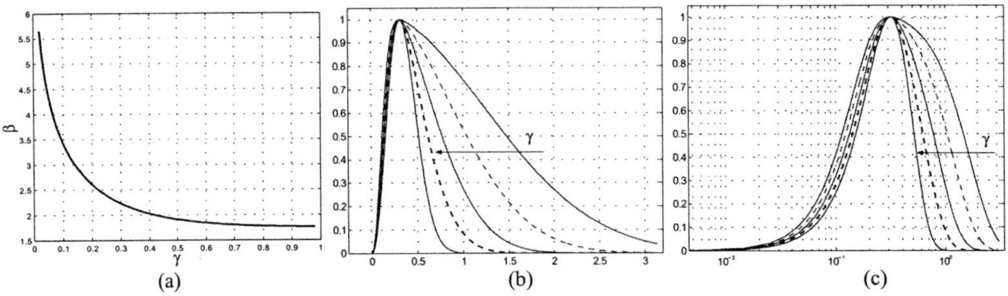

Figure 4: Difference of Gaussian filter. Bandwidth as a function of γ (a); Fourier transform of the filter on a linear (b) and logarithmic (c) scale for $\lambda_0 = 20$ and $(\gamma, \beta) = (0.081\ , 3.63)$, $(0.131\ , 3.05)$, $(0.212\ , 2.53)$, $(0.343\ , 2.12)$ and $(0.9\ , 1.76)$. When $\gamma \approx 0.9$ the filter is a good approximation to the 2nd derivative of a Gaussian.

Therefore, the Fourier transform of the quadrature pair is defined by (up to a multiplicative factor of 2):

$$F_{DoG}(\omega) = \begin{cases} n_c \exp\left(-\frac{\sigma'^2\omega^2}{2}\right) - n_c \exp\left(-\frac{\sigma^2\omega^2}{2}\right) \ , & \text{if } \omega \geq 0 \ , \\ 0 \ , \quad \text{otherwise} \ . \end{cases} \tag{39}$$

and the peak tuning frequency is given by:

$$\omega_0 = \frac{2}{\sigma}\left(\frac{\ln(\gamma)}{(\gamma^2-1)}\right)^{1/2} \quad \text{where,} \quad \gamma = \frac{\sigma'}{\sigma} \ . \tag{40}$$

The two values for the normalisation constant n_c for this family are given by:

$$\max_{\omega} F_{DoG}(\omega) = 1 \quad \Rightarrow n_c = 1/F_{DoG}(\omega_0) \ , \tag{41}$$

$$\|f_{DoG}(x)\| = \frac{1}{2\,\pi}\|F_{DoG}(\omega)\| = 1 \quad \Rightarrow n_c = 2\sqrt{\sqrt{\pi}\,\sigma}\left(\frac{1}{\gamma} - \frac{2\sqrt{2}}{\sqrt{1+\gamma^2}} + 1\right)^{-\frac{1}{2}} \ . \tag{42}$$

The quadrature DoG filter is invariant to grey level shift (zero-DC) and invariant to an additive ramp (use eq. (27) with eq. (39)).

Unfortunately, we have not been able to derive an analytical expression for the bandwidth of the filter. It's straightforward to note that the bandwidth depends only on the ratio $\gamma = \sigma'/\sigma$. The bandwidth as a function of γ is shown in Fig. 4 (a). Notice that the smallest bandwidth allowable for this family is 1.76 (corresponding to the bandwidth of the 2nd derivative of a Gaussian). The filters for $\lambda_0 = 20$ and for different values of γ are shown in Fig. 4 (b) and (c) on a linear and logarithmic scale respectively. Observe that the tuning frequency must be chosen carefully to avoid aliasing problem especially for small values of γ.

2.5 Cauchy Filter

Cauchy filters have been proposed in [47, 78, 79]. This family does not appear to be well known in the computer vision community. However, this family has some attractive properties as we see below. The Cauchy functions are given by:

$$f_n(x) = \Re\{(1-ix)^{-n-1}\} = \frac{\Re\{(1+ix)^{n+1}\}}{(1+x^2)^{n+1}} \quad ; \tag{43}$$

$$g_n(x) = \Im\{(1-ix)^{-n-1}\} = \frac{\Im\{(1+ix)^{n+1}\}}{(1+x^2)^{n+1}} \quad ; \tag{44}$$

where $n \in \mathbb{N}$ and determines the number of ripples of the two functions and $\Re(z)/\Im(z)$ is the real/imaginary part of z. f_n and g_n are in quadrature and are integrable for $n \geq 1$. The Fourier transform of f_n is given by [79]:

$$F_n(\omega) = 2\,\pi(n!)^{-1}|\omega|^n \exp(-|\omega|) \ . \tag{45}$$

The values $n = 3$ and $n = 5$ have been proposed to model some properties of the human visual response in [47].

Following the same procedure as for the Gaussian derivatives, we can define a family of quadrature filters using Cauchy functions for $(n = a \geq 1) \in \mathbb{R}$; which in the Fourier domain gives:

$$F_{Ch}(\omega) = \begin{cases} n_c\,\omega^a \exp(-\sigma\,\omega) & \text{if} \quad \omega \geq 0 \ , \\ 0 & \text{otherwise} \ . \end{cases} \tag{46}$$

where σ is a scaling parameter. Hence, by definition, the filters have zero-DC and are invariant to an additive ramp for $a > 1$. Notice that the expression of the filters is similar to that for the Gaussian derivative filters. Actually, the restriction of eq. (46) to $a \in \mathbb{N}$ and subject to the appropriate normalisation, is the Erlang distribution. The Erlang functions are well known in statistics [see eg. 71] and they have been used recently for the design of a 2D quadrature filters, steerable in scale and orientation in [3]. The reader is referred to [3] for further properties of this family when used in a multiscale framework.

It can be shown that the filters in the spatial domain are given by:

$$f_{Che}(x) = n_c \frac{\Gamma(a+1)}{2\,\pi\,\Re(\sigma - ix)^{a+1}} = n_c \frac{\Gamma(a+1)}{2\,\pi} \frac{\Re(\sigma+ix)^{a+1}}{(\sigma^2+x^2)^{a+1}} \ , \tag{47}$$

$$f_{Cho}(x) = n_c \frac{\Gamma(a+1)}{2\,\pi\,\Im(\sigma - ix)^{a+1}} = n_c \frac{\Gamma(a+1)}{2\,\pi} \frac{\Im(\sigma+ix)^{a+1}}{(\sigma^2+x^2)^{a+1}} \ , \tag{48}$$

and the normalisation constants are given by:

$$\max_{\omega} F_{Ch}(\omega) = 1 \quad \Rightarrow n_c = 1/F_{Ch}(\omega_0) \ , \tag{49}$$

$$\|f_{Ch}(x)\| = \frac{1}{2\,\pi}\|F_{Ch}(\omega)\| = 1 \quad \Rightarrow n_c = \left(\frac{\pi\,4^{(a+1)}\,\sigma^{(2\,a+1)}}{\Gamma(2\,a+1)}\right)^{\frac{1}{2}} \ , \tag{50}$$

where ω_0 is the peak tuning frequency of the filter and is given by the following ratio:

$$\omega_0 = \frac{a}{\sigma} \ . \tag{51}$$

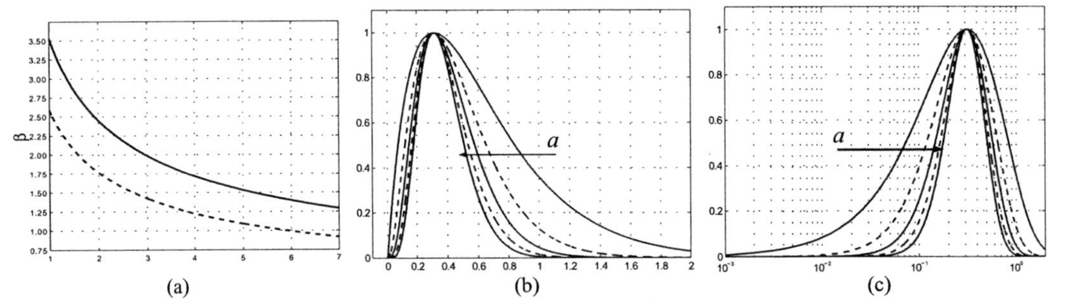

Figure 5: Cauchy quadrature filters; Bandwidth β as a function of a; Cauchy filters (solid curve) and Gaussian derivative filter (dashed curve) (a). Fourier transform of the quadrature pair on a linear (b) and logarithmic (c) scale for $a = 1, 2, 3, 4$ and 5.

The octave bandwidth of this family is given by:

$$\beta = \frac{\ln\left(\frac{\mathrm{W}(-1,\, c)}{\mathrm{W}(0,\, c)}\right)}{\ln(2)} \,, \qquad \text{with} \quad c = -\frac{2^{-(1/a)}}{e} \,. \tag{52}$$

Finally, the complex scalar $z(\Delta x,\, \Delta\lambda)$ which encodes the phase behaviour of this family for a small spatial perturbation $\Delta\, x$, and a small scale perturbation $\Delta\, \lambda$ is given by:

$$z_{Ch}(\Delta x,\, \Delta\lambda) = \left(\frac{2\sqrt{\sigma_1\,\sigma_0}}{\sigma_1 + \sigma_0 - i\,\Delta x}\right)^{2\,a+1} = \left(\frac{2\,a\,\sqrt{\lambda_1\,\lambda_0}}{a\,(\lambda_1 + \lambda_0) - i\,2\pi\,\Delta x}\right)^{2\,a+1} \,. \tag{53}$$

Figure 5 (a) shows the evolution of the bandwidth of the filters function of a. The bandwidth of the Gaussian derivatives is shown for comparison. The transfer function of the quadrature pair is shown on a linear and logarithmic scale in Fig. 5 (b)-(c).

2.6 Quadrature Filters Related to Canny's Criteria

Several papers have considered designing optimal edge/line detectors in the sense of Canny's criteria [9] (see Appendix B). These detectors are typically the first/2nd derivative of a smoothing function. To our knowledge, two smoothing functions have been proposed in this sense.

2.6.1 Deriche Filter

We can define a family of quadrature filters using the smoothing function proposed by Deriche [16]. This function is the integral of the optimal operator in the sense of Canny's criteria for edge detection [9]. It is also the smoothing function used by Ziou [95] for optimal line detection according to Canny's criteria. Hence, it is the second integral of the optimal line detector and is given by:

$$s_d(x) = (\alpha\,|x| + 1)\,\exp(-\alpha\,|x|) \,, \tag{54}$$

$$S_d(\omega) = \frac{4\,\alpha^3}{(\omega^2 + \alpha^2)^2} \,, \tag{55}$$

Table 1: Coefficients of the fitting function given by eq. (60).

i	c_i	μ_i
0	2.54372940760	–
1	0.11940325096	2.00000000000
2	0.01269121401	0.58167860542

where α is the scaling parameter ($\sim \sigma^{-1}$).

In a similar way as for Gaussian derivative filters we define a family of quadrature filters in the Fourier domain:

$$F_d(\omega) = \begin{cases} n_c\,\omega^a\,S_d(\omega)\,, & \text{if}\,\omega \geq 0\,; \\ 0\,, & \text{otherwise}\,. \end{cases} \qquad \text{where,} \quad 1 \leq a \leq 2\ . \tag{56}$$

The odd filter is the Deriche edge detector for $a = 1$ and the even filter is Ziou's line detector when $a = 2$. The filters are invariant to a grey level shift, and invariant to an additive ramp for $a > 1$ (see eq. (26)).

It can be shown that the peak tuning frequency is the following function:

$$\omega_0 = \alpha \sqrt{\frac{a}{4-a}}\ , \tag{57}$$

and, the two values for the normalisation constant n_c are given by:

$$\max_{\omega} F_d(\omega) = 1 \quad \Rightarrow n_c = 1/F_d(\omega_0)\ , \tag{58}$$

$$\|f_d(x)\| = \frac{1}{2\,\pi}\|F_d(\omega)\| = 1 \quad \Rightarrow n_c = \alpha^{(\frac{1}{2}-a)} \sqrt{\frac{-12\,\cos(\pi\,a)}{(2\,a-5)(2\,a-3)(2\,a-1)}}\ . \tag{59}$$

Unfortunately, there is not a simple analytical expression for the bandwidth of the filter. We did a numerical estimation of the bandwidth and approximated its functional dependence on a by the following 4th order polynomial:

$$\beta = c_0 + c_1 \left(\frac{a-\mu_1}{\mu_2}\right)^2 + c_2 \left(\frac{a-\mu_1}{\mu_2}\right)^4\ . \tag{60}$$

The coefficients are tabulated in Table 1. The maximum relative error with respect the numerical estimation is 0.0006.

2.6.2 Sarkar and Boyer Filter

Sarkar and Boyer [80] proposed a modification of the multiple response criteria appropriated to the use of IIR filters. Following a similar approach as Canny [9] and Deriche [16], the authors proposed the following smoothing function:

$$s_s(x) = \frac{1}{\alpha}\,\exp(-\alpha\,|x|)\left\{\cos(\varphi) - \frac{\cos(\alpha\,\tilde{m}\,|x| + \varphi) - \tilde{m}\,\sin(\alpha\,\tilde{m}\,|x| + \varphi)}{\tilde{m}^2+1}\right\}\ , \tag{61}$$

Table 2: Bandwidth and peak tuning frequency of $F_s(\omega)$ for some particular values of the shape parameters. $\Sigma\Gamma$ and Ω_S are given when the odd filter is used for edge detection alone. The Gaussian derivative filter is shown for comparison. (*) Here the filter is equivalent to the Deriche filter, $\tilde{m} = 0.002$ is the smallest value in our sampling space. Basically the true values for the Deriche filters should be $\tilde{m} = 0$ and $\varphi = \frac{\pi}{2}$.

	$\Sigma\Gamma$	Ω_S	$\tilde{m}$	φ	ω_0/α	β	
	0.885	3.252	1.917	0.651	1.354	2.700	
$a = 1$	2.000	1.622	0.002	1.382	0.577	3.009	(*)
	1.833	1.833	0.074	0.120	0.491	2.933	
	1.443	2.453	0.464	0.332	0.535	2.787	
GD	0.921	3.245	-	-	$\omega_0 = 1/\sigma$	2.590	
	-	-	1.917	0.651	1.746	1.555	
$a = 2$	-	-	0.002	1.382	1.000	2.543	(*)
	-	-	0.074	0.120	0.814	2.401	
	-	-	0.464	0.332	0.823	2.139	
GD	-	-	-	-	$\omega_0 = \sqrt{2}/\sigma$	1.764	

where φ and $\tilde{m}$ determine the shape of the filter and α is the scaling parameter. Note that the above smoothing function is equivalent to eq. (54) for $\varphi = \frac{\pi}{2}$ and when $\tilde{m} \to 0$.

The Fourier transform of (61) is:

$$S_s(\omega) = 2\,\alpha^2\,\tilde{m}\,\frac{(2\,\sin(\varphi) - \tilde{m}\,\cos(\varphi))\,\omega^2 + \alpha^2\,(\tilde{m}^3\,\cos(\varphi) + 3\,\tilde{m}\,\cos(\varphi) + 2\,\sin(\varphi))}{(\omega^2+\alpha^2)((\omega+\alpha\,\tilde{m})^2+\alpha^2)((\omega-\alpha\,\tilde{m})^2+\alpha^2)}\,, \tag{62}$$

Hence, we can define the following family of quadrature filters in the Fourier domain:

$$F_s(\omega) = \begin{cases} n_c\,\omega^a\,S_s(\omega)\,, & \text{if}\,\omega \geq 0\,; \\ 0\,, & \text{otherwise}\,. \end{cases} \qquad \text{where,} \quad 1 \leq a \leq 2\;. \tag{63}$$

Unfortunately, the Fourier representation of the filters is not easy to manipulate analytically. Therefore, we studied the filters numerically for $a \in [1, 2]$, and for 4 combinations of the shape parameters $(\tilde{m}, \varphi)$ corresponding to the best values of $\Omega_S(fs)$, $\Sigma\Gamma(fs)$, $\Sigma\Gamma = \Omega_S(fs)$ and $\Sigma\Gamma\Omega_S(fs)$ for the odd part when $a = 1$ (i.e Sarkar's edge operator). The values are tabulated in Table 2. Figure 6 (a) shows the evolution of the bandwidths as a function of a. We first notice a high correlation between the SNR-localisation and the bandwidth of the filters; $\Sigma\Gamma$ increases when β increases. The 1st derivative of the Gaussian is an exception to this rule. We believe that this is mainly due to the difference in the decay of the Gaussian function (exponential) and Sarkar's smoothing function (ω^{a-4}). This is illustrated in Fig. 6 (c) and (d). Values of ω_0/α give some insight regarding the spatial support of the filters (see Fig. 6 (b)).

We might expect that the performance of a detector depends on its spatial shape. However, this experiment shows the importance of the bandwidth of the filter. Notice that the smallest bandwidth for the edge operators is 2.6 octaves. Several authors mention that larger

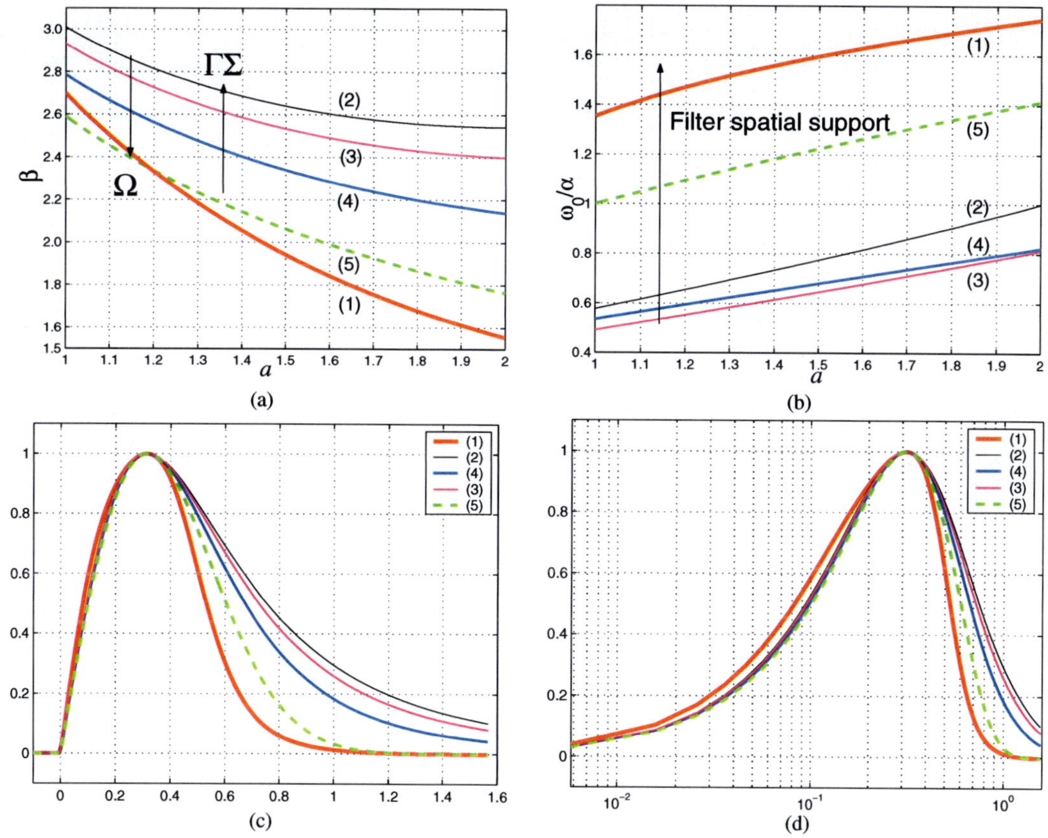

Figure 6: Sarkar and the Deriche filters. Bandwidth as a function of a (a); ω_0/α as a function of a (b); Filters in the Fourier domain on a linear (c) and logarithmic (d) scale for $\lambda_0 = 20$ and $a = 1$. best Ω_S (curves (1)); best $\Sigma\Gamma$ (curves (2)); $\Omega_S = \Sigma\Gamma$ (curves (3)); best $\Sigma\Gamma\Omega_S$ (curves (4)) and Gaussian derivative filter (curves (5)). The numbers refer to the same order in Table 2.

bandwidths give better results in phase based feature detection [36, 49, 68]. A bandwidth of 1.5 octaves is used in [36, 37, 46, 77], 1.76 octaves in [56, 76] and 2 octaves in [49, 68]; all smaller than 2.6 and even smaller then 2.54 (bandwidth of Ziou's line detector). The second important thing that we can learn from these values, is that even the smallest bandwidth for the edge operators is larger than the largest one for line detectors. Knowing for example that, for the Deriche filter, the odd part alone, for $a = 1$, is optimal in the sense of Canny for edge detection and the even part, for $a = 2$, is optimal in the same sense for lines [8], this implies that a quadrature pair will not be optimal for both in the sense of Canny. Hence, a tradeoff must be made.

The spatial representation of the filters is shown in Fig. 7. Figure 7(a)-(b) shows the filters for $a = 1$ (i.e odd parts correspond to the edge operators) and Fig. 7(c)-(d) shows the same filters but for a constant bandwidth of 2.543 octaves (i.e different values of a).

[8]Of course the optimality is debatable as for example we found that the even DoG filter (when $\gamma \leq 0.05$) has better performance than Ziou's line detector, in the sense of Canny, using the following line model: $F(x) = \mathrm{rect}((x - 0.25)/0.5)$.

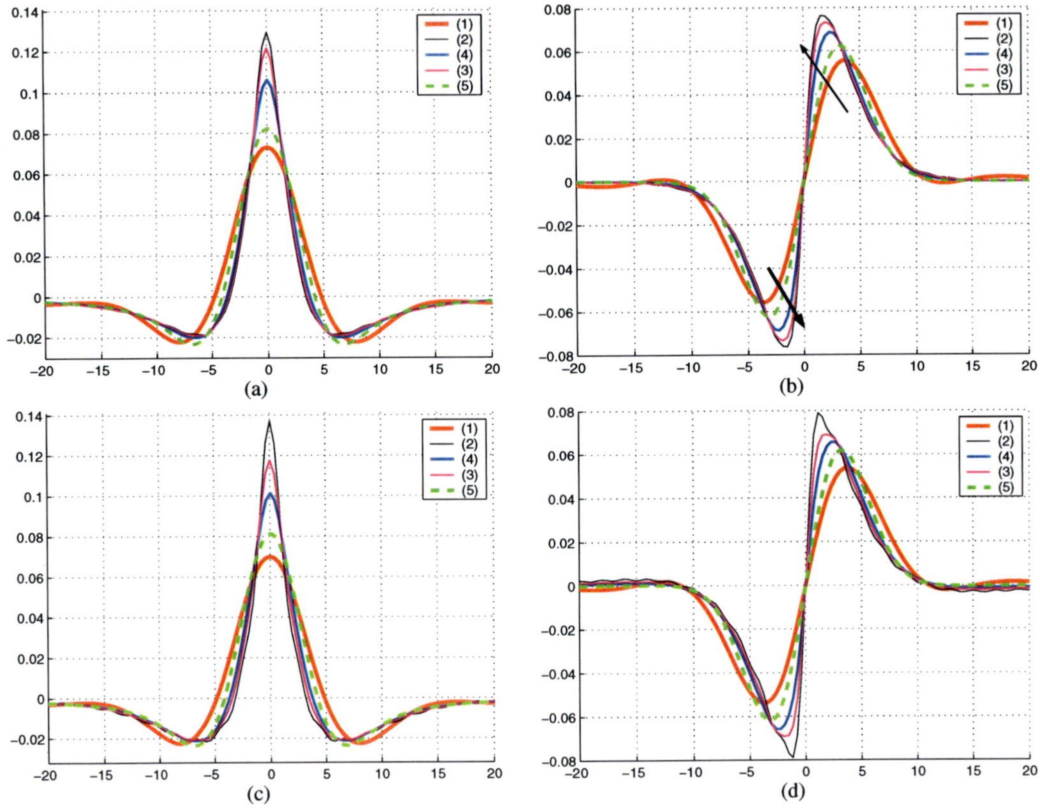

Figure 7: Sarkar and the Deriche filters in the spatial domain for $\lambda_0 = 20$. Constant value of $a = 1$ (a) and (b), the arrows show increasing $\Sigma\Gamma$ (see Fig. 6 (c)-(d) for the Fourier representation); Constant bandwidth $\beta = 2.543$ octaves (c) and (d). Best Ω_S (curves (1)); best $\Sigma\Gamma$ (curves (2)); $\Omega_S = \Sigma\Gamma$ (curves (3)); best $\Sigma\Gamma\Omega_S$ (curves (4)) and Gaussian derivative filter (curves (5)).

Interestingly, the shape (by which we mean the relative differences) of the filters are similar for both cases. Hence we expect the filters to have similar relative performances in the sense of Canny. This experiment shows the importance of filter shape. Notice that a bandwidth of 2.543 corresponds to $a = 2$ for the Deriche filter (optimal even part) and to $a = 1.032$ for the Gaussian derivatives (good approximation to the 1st derivative). Inspecting Fig. 7(d), it is more likely that the odd part of a Deriche filter for $a = 2$ will have a better $\Sigma\Gamma$ than the 1st derivative of a Gaussian.

The above experiments highlight issues in choosing a quadrature pair for feature detection and on which basis. It also suggests the possibility of adapting Canny's criteria to the local energy model for the purpose of filter design aiming to have similar performances at least for edges and lines. This is a very delicate point since, for example, the localisation criteria and the multiple responses criteria might not be suitable as they do not take into account the phase information. However, the adaptation of the SNR criteria is straightforward.

We do not have a clear answer to the above questions but they are worthy of further investigation. It is more likely and justified, however, to use a quadrature pair where the odd/even part is optimal in the sense of Canny if we are more interested in detecting edges/lines, assuming that the contribution of the even/odd part of the filter is negligible.

Table 3: Summary of main equations and filters properties.

	Spatial domain	Fourier domain	Bandwidth	Tuning frequency	zero-DC /ramp invariance
Gabor	$n_c \frac{e^{-\frac{x^2}{2\sigma^2}}}{\sqrt{2\pi}\,\sigma}\, e^{i\,\omega_0\, x}$	$n_c\, e^{-\frac{\sigma^2}{2}(\omega-\omega_0)^2}$	$\frac{\ln\left(\frac{\sigma\,\omega_0+\sqrt{2\,\ln(2)}}{\sigma\,\omega_0-\sqrt{2\,\ln(2)}}\right)}{\ln(2)}$	ω_0	No / No
log-Gabor	No	$n_c\, \exp\left(-\frac{\ln^2(\omega/\omega_0)}{2\,\ln^2(\kappa_\beta)}\right)$	$-\frac{2\sqrt{2}}{\sqrt{\ln 2}}\,\ln(\kappa_\beta)$	ω_0	Yes/Yes
Gaussian Derivative	Only for $a \in \mathbb{N}$	$n_c\,\omega^a\, \exp[-\sigma^2\omega^2]$	$\frac{\ln\left(\frac{\mathrm{W}(-1,\, c)}{\mathrm{W}(0,\, c)}\right)}{2\,\ln(2)},\ c=-\frac{4^{-\frac{1}{a}}}{e}$	$\frac{\sqrt{a}}{\sigma}$	Yes/Yes ($a > 1$)
DoG	Yes	$n_c(e^{-\frac{\sigma'^2\,\omega^2}{2}} - e^{-\frac{\sigma^2\,\omega^2}{2}})$	Numerical Fig. 4 (a)	$\frac{2}{\sigma}\sqrt{\frac{\ln(\gamma)}{(\gamma^2-1)}},\ \gamma=\frac{\sigma'}{\sigma}$	Yes/Yes
Cauchy	$n_c\,\frac{\Gamma(a+1)}{2\,\pi\,(\sigma-ix)^{a+1}}$	$n_c\,\omega^a\,\exp(-\sigma\,\omega)$	$\frac{\ln\left(\frac{\mathrm{W}(-1,\, c)}{\mathrm{W}(0,\, c)}\right)}{\ln(2)},\ c=-\frac{2^{-\frac{1}{a}}}{e}$	$\frac{a}{\sigma}$	Yes/Yes ($a > 1$)
Deriche	No	$n_c\,\omega^a\,\frac{4\,\alpha^3}{(\omega^2+\alpha^2)^2}$	Numerical Fig. 6 (a)	$\alpha\,\sqrt{\frac{a}{4-a}}$	Yes/Yes ($a > 1$)
Sarkar & Boyer	No	Yes (*)	Numerical Fig. 6 (a)	Numerical Fig. 6 (b)	Yes/Yes ($a > 1$)

(*) $2\,n_c\,\omega^a\alpha^2\,\tilde{m}\,\frac{(2\,\sin(\varphi)-\tilde{m}\,\cos(\varphi))\,\omega^2+\alpha^2\,(\tilde{m}^3\,\cos(\varphi)+3\,\tilde{m}\,\cos(\varphi)+2\,\sin(\varphi))}{(\omega^2+\alpha^2)((\omega+\alpha\,\tilde{m})^2+\alpha^2)((\omega-\alpha\,\tilde{m})^2+\alpha^2)}$

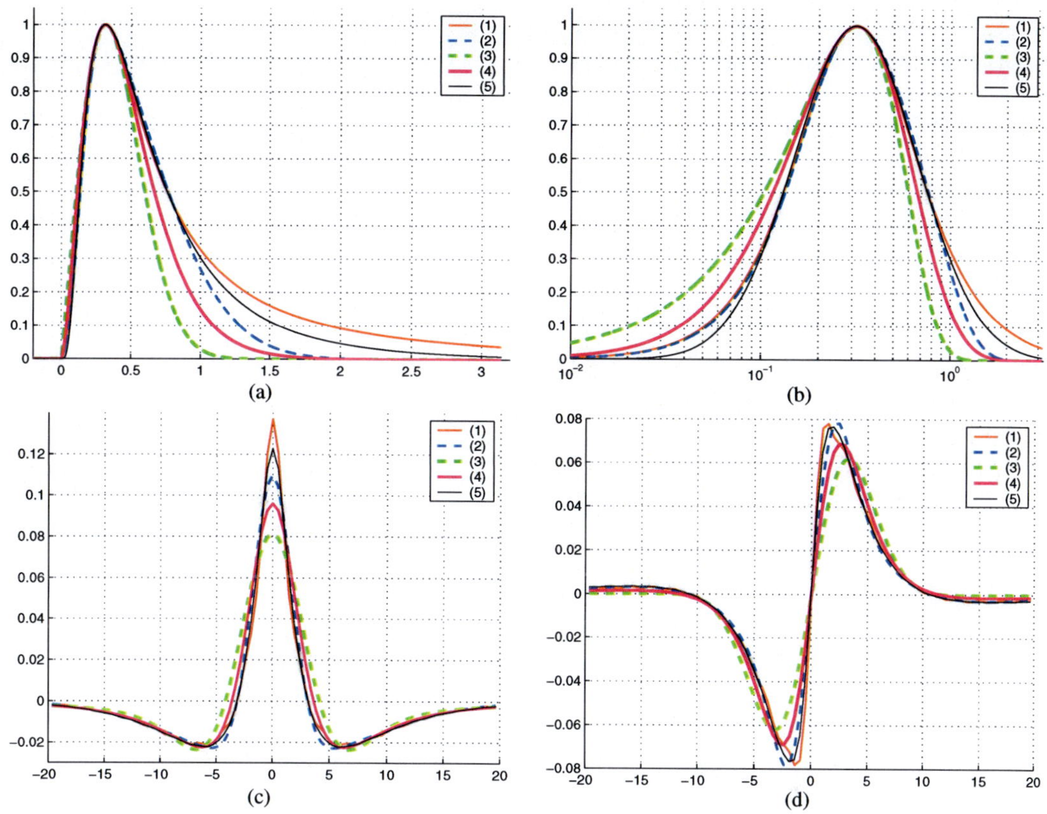

Figure 8: Different quadrature filters in the frequency domain (a) and (b) and in the spatial domain, even part (c) and odd part (d). All the filters have a bandwidth of $\beta = 2.543$ octaves and a peak tuning frequency for $\lambda_0 = 20$. Deriche filter with $a = 2$ (curves (1)); DoG filter with $\gamma = 0.215$ (curves (2)); Gaussian derivative with $a = 1.032$ (curves (3)); Cauchy filters with $a = 1.859$ (curves (4)) and log-Gabor filters with $\kappa_\beta = 0.473$ (curves (5)).

3 Comparison

We have presented most of the mathematical properties of the different families in the previous section. A summary of the key equations and properties is given in Table 3. In this section, we present a comparison of the filters used as a quadrature pair. For this purpose, the comparison will be fair only if the filters have the same bandwidth and the same tuning frequency.

Figure 8 shows the five different families of filters (Deriche, DoG, Gaussian Derivative, Cauchy and log-Gabor) for a bandwidth of 2.543 octaves (Bandwidth of Ziou's line detector) and a wavelength of 20, both in the frequency domain and in the spatial domain. Inspecting the Fourier representation of the filters, we notice that the Gaussian derivative filters have the least aliasing, and the Deriche filter is the worst. Notice also that the three families, Deriche, log-Gabor and the DoG filter are very similar for this bandwidth, and the Cauchy family is between these three filters and the Gaussian derivative filter. This gives

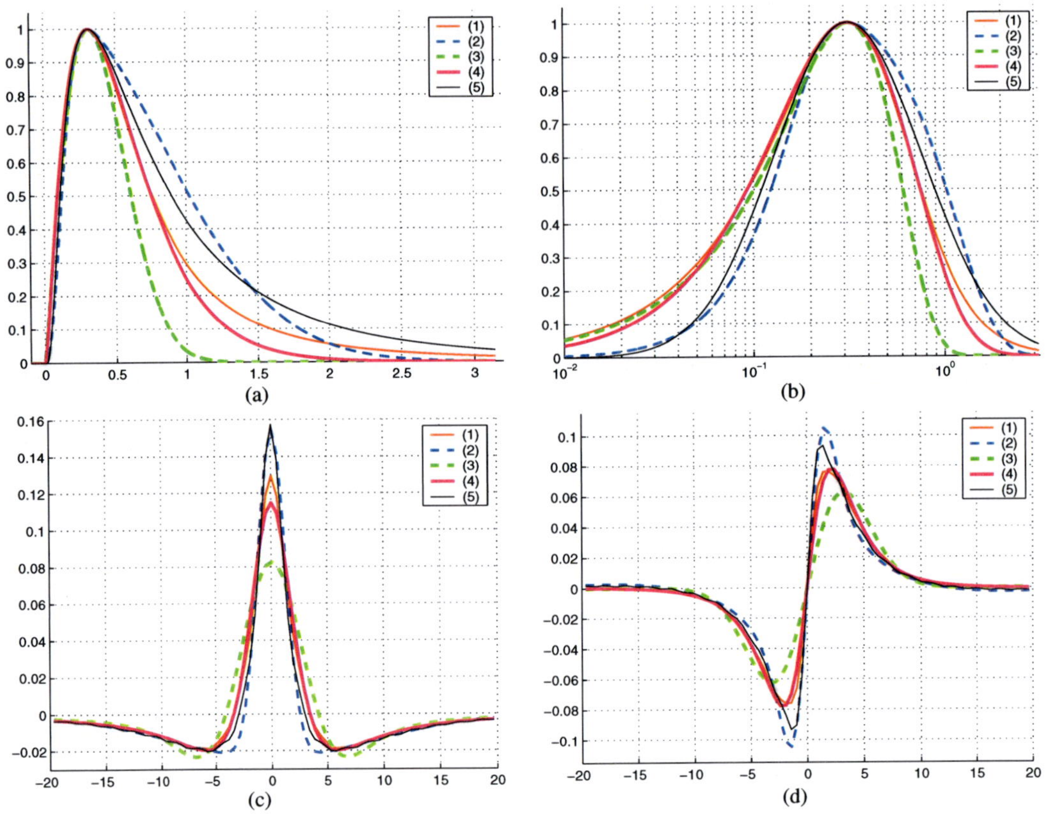

Figure 9: Quadrature filters in the frequency domain (a) and (b) in the spatial domain, even part (c) and odd part (d). All the filters have a bandwidth of $\beta = 3.009$ octaves (except Gaussian derivative $\beta = 2.590$) and a peak tuning frequency for $\lambda_0 = 20$. Deriche filter with $a = 1$ (curves (1)); DoG filter with $\gamma = 0.135$ (curves (2)); Gaussian derivative with $a = 1$ (curves (3)); Cauchy filters with $a = 1.348$ (curves (4)) and log-Gabor filters with $\kappa_\beta = 0.412$ (curves (5)).

us some insight into the behaviour of the filters for feature detection as we know that the even part of the Deriche filter is optimal in the sense of Canny, and the odd part of the Gaussian has a high Ω_S and a low $\Sigma\Gamma$. The log-Gabor filter is an acceptable approximation of Ziou's line detector (even part of the Deriche filter), and has low aliasing. Hence, if we use a Fourier domain implementation, this suggests that the log-Gabor filter is preferable for line detection using this bandwidth.

Figure 9 shows the same five families for the same wavelength of 20 but for a larger bandwidth ($\beta = 3.009$ octaves) in the frequency domain and in the spatial domain, except for the Gaussian derivative filter where $\beta = 2.590$ (maximum bandwidth corresponding to the 1st derivative). A bandwidth of 3.009 octaves corresponds to the Deriche edge operator for the Deriche family which has the best SNR-localisation. First of all, notice that for this bandwidth, the Deriche filter has less aliasing than the log-Gabor filter. This is the opposite relation to before (i.e, when $\beta = 2.543$ octaves). We remark that the Cauchy filter approximates the Deriche filter quite well and has the advantage of being invariant to an

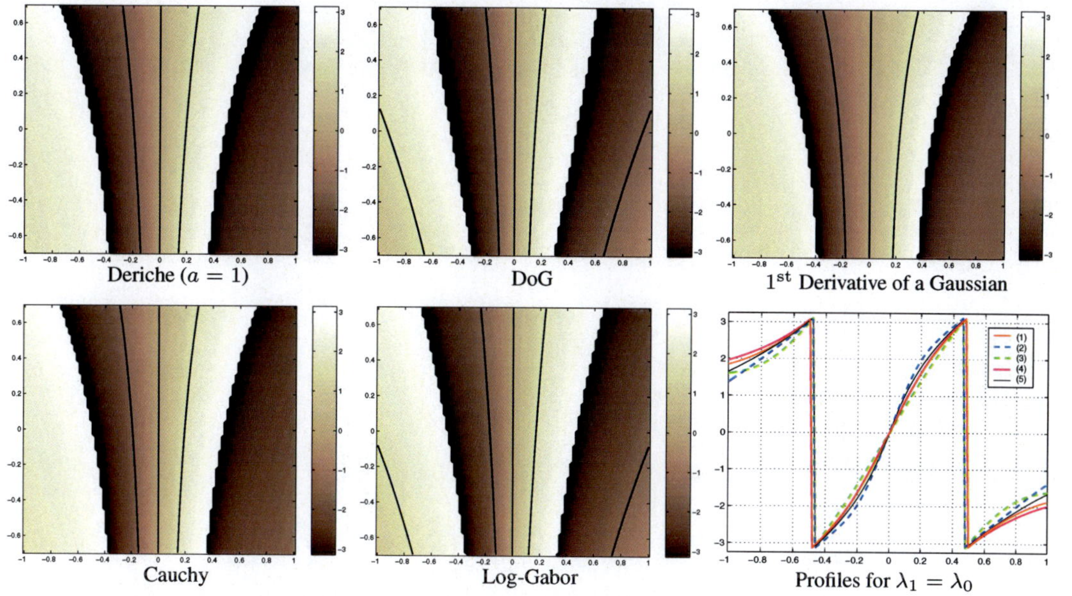

Figure 10: Mean phase difference level contours $\mu(z) = n\,\pi/2$, with $n = -4, \ldots, 4$, near $\mathbf{p}_0 = (0, \lambda_0)$ with a log scale on the vertical axis over two octaves $\frac{\lambda_1}{\lambda_0} \in [\frac{1}{2}, 2]$, and spatial position on the horizontal axis, $\frac{x_1}{\lambda_0} \in [-1, 1]$. All the filters have a bandwidth of 3.009 octaves except for the 1st derivative of a Gaussian.

additive ramp besides the fact that Cauchy filters has less aliasing than the Deriche filter. The DoG filter seems to be very similar to the log-Gabor filter except that the DoG filter has less aliasing. Notice that, the DoG filter with this bandwidth (i.e $\gamma = 0.135$) has a better SNR-localisation than Ziou's optimal line detector (even part of the Deriche filter for $a = 2$) and similar performance in term of the product of the three criteria.

3.1 Phase Stability

Phase stability is very important in phase-based estimation techniques. Stability over scales is important for example in feature detection using Phase Congruency and the linearity of phase with respect to a space perturbation is more important in velocity estimation. It is also desirable to know the error in measurement of phase due to the presence of noise. This could for example be taken into account in a regularisation framework of the estimation/detection.

Fleet and Jepson [29] studied the phase behaviour of quadrature filters with respect to a scale-space perturbation $(\Delta x, \Delta\lambda)$ in the neighbourhood of a scale-space point $\mathbf{p}_0(0, \lambda_0)$ in the presence of white Gaussian noise. A short summary of their main results is given in Appendix A. The authors derived an approximation to the expected mean $\mu(z)$ of phase difference $\Delta\phi$, and an upper bound of the expected mean variation $b(z)$ about the mean phase difference $\mu(z)$. $\mu(z)$ is independent of the bandwidth of the filters but $b(z)$ is not. Phase is stable when $\Delta\phi$ is constant with respect to changes in scale $\Delta\lambda$; and it is linear with respect to spatial position when $\Delta\phi$ is a linear function of Δx (i.e stability through scale and linearity through space near $\mathbf{p}_0$). Earlier, in this chapter, we provided analytical expressions for phase behaviour for the Gabor and Cauchy kernels, and we studied the

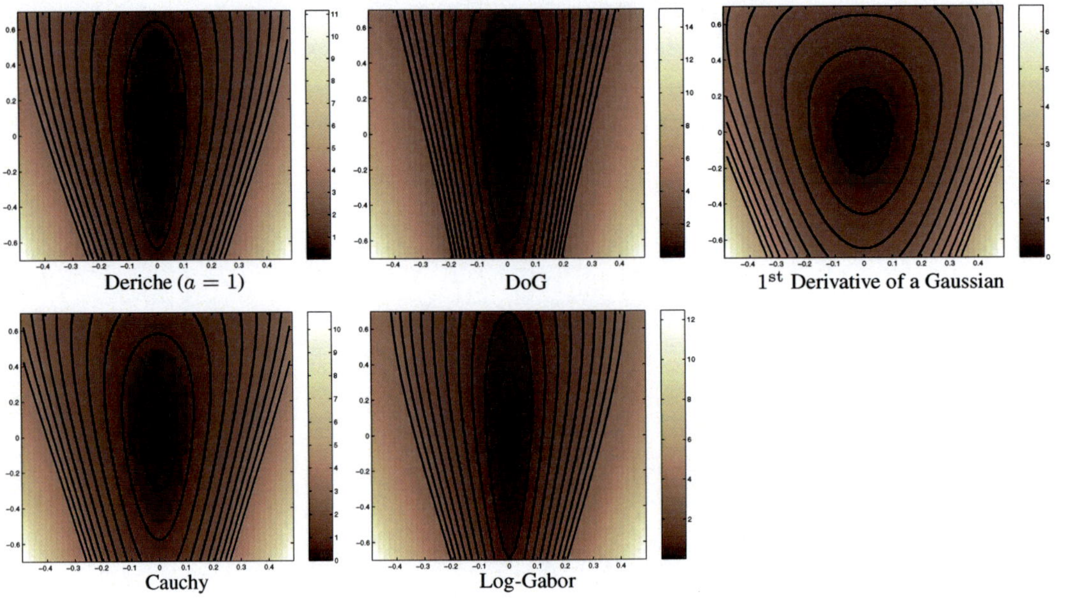

Figure 11: Level contours $b(z) = 0.3\,n$, with $n = 1, \ldots, 10$, corresponding to $\mu(z)$ shown in Fig. 10. All the filters have a bandwidth of 3.009 octaves except for the 1st derivative of a Gaussian.

others numerically.

Figure 10 shows the level contours of the mean phase difference $\mu(z)$, for five families: Deriche, DoG, Cauchy and log-Gabor for $\beta = 3.009$ and the 1st derivative of a Gaussian for $\beta = 2.590$. In theory $\mu(z)$ is independent of the bandwidth of the filters, if the self-similarity condition A.7 is satisfied. This figure illustrates that phase is stable over scale for all the families. However the linearity through space is not perfect for some of them for this bandwidth. This is illustrated in Fig. 10 (Profiles), showing $\mu(z)$ as a function of Δx when $\lambda_1 = \lambda_0$.

We find that phase is perfectly linear only for the 1st derivative of a Gaussian when $-\frac{\lambda_0}{2} \leq x_1 \leq \frac{\lambda_0}{2}$; Cauchy filters and the Deriche filter have a similar behaviour and are close to linear in this range. However the phase is still linear in the vicinity of $\mathbf{p}_0$ for all filters. The DoG and log-Gabor kernels have the largest phase distortion. This is a consequence of the tailed nature of the two families for large bandwidths. Inspecting the errors on the mean estimation shown in Fig. 11, we notice that it is for these two families that we have the largest error bound for a spatial perturbation and the smallest for a scale perturbation. This observation, as well, is a consequence of the tailed nature of the log-Gabor and the DoG kernels. As mentioned by Fleet and Jepson [29], as the bandwidth increases, the spatial span of the kernels decreases, and therefore there is a decrease in the spatial extent over which phase is linear. On the other hand, an increase in bandwidth implies that filters tuned to nearby scales will overlap to a great extent, and therefore phase is stable for larger scale perturbations. The 1st derivative of a Gaussian has a smaller bandwidth in this experiment, hence, phase is linear for a larger extent. Regarding the other kernels, the larger the extent of the kernel in the Fourier domain, the less is the extent of phase linearity.

Notice that there is a trade-off for the bandwidth between the scale perturbation on the mean phase error $b(z)$, and the spatial one. We will come back to this point when analysing stability for small bandwidths below.

Figure 12, shows the same measures $\mu(z)$ and $b(z)$ for a smaller bandwidth $\beta = 2.543$ octaves (corresponds to the bandwidth of Ziou's line detector). We notice the same observations regarding the error on the expected mean phase as before. Notice that as the bandwidth of the filters decreases, the expected deviation from $\mu(z)$ is reduced for spatial perturbations and increases for scale perturbations. This is not the case for the Deriche filter, for which the effect is slightly the opposite (see for example the difference between the DoG filter and the Deriche filter). Indeed, as we already mentioned in the last section (see Fig. 8), the Deriche filter has the largest aliasing for this bandwidth (larger than for $\beta = 3.006$, i.e $a = 1$). This is due to the decay of the Fourier transform of this family ($\omega^{(a-4)}$). Regarding the phase linearity of the filters, as expected, we notice that the extent over which phase is linear has increased.

The phase behaviour for a bandwidth of $\beta = 1.8$ octaves is shown in Fig. 13 for: Gabor kernels with zero-DC correction as proposed in (Appendix C), DoG, derivative of a Gaussian, Cauchy and log-Gabor kernels. We did not show the Deriche filters result as the smallest bandwidth of this family is 2.54 octaves. For $\beta = 1.8$, all the filters are perfectly linear for spatial perturbations of more than half of a wavelength. Notice the similar behaviour in terms of the expected deviation $b(z)$ for Gabor, DoG and the Gaussian derivative filters. Notice that a Gaussian derivative filter of a bandwidth of 1.8 octaves corresponds to $a = 1.92$ and a DoG filter of the same bandwidth has a value of 0.7 for γ. Hence, the similarity between the DoG filter and the derivative of a Gaussian filter is expected for this bandwidth, as we know that the 2nd derivative of a Gaussian is well approximated by the DoG filter for $\gamma \approx 0.9$ and that the shape variability of the DoG filter is very low when $\gamma > 0.5$ (see Fig. 4).

Finally, Fig. 14 shows $\mu(z)$ and $b(z)$ for a bandwidth of $\beta = 1$ octave for: Gabor kernels with (proposed in Appendix C) and [78]) as zero-DC correction, DoG, derivative of a Gaussian, Cauchy and log-Gabor filters. All the families have a linear phase over the range of x_1. All the filters have a comparable errors on the estimation of the mean phase. Notice, the circular shape of the level contours of $b(z)$. As we mentioned before, for small bandwidths, the expected deviation from the mean phase for spatial perturbations of the input is reduced and there is an increase of the spatial extent over which phase is linear. On the other hand, there is an increase of the error bound for scale perturbations which decreases the extent of phase stability.

3.2 Uncertainty Principle

The uncertainty principle criteria is a measure of the scale-space localisation of the filters. Gabor functions are known to be optimal in this sense [see e.g. 60, Chap. 2]. Figure 15 shows the standard deviations of the energy of the filters in the spatial domain σ_f and in the frequency domain σ_F and the uncertainty principle, which is the product of both, $\sigma_f\,\sigma_F$; all as a function of the bandwidth of the filters and for a peak tuning frequency for $\lambda_0 = 20$. σ_f and σ_F are dependent on λ_0 but not the product. The families used in this experiment are: Cauchy filters (1), Gabor with Ronse zero-DC correction using $\tilde{\sigma} = \sigma$ (2) and $\tilde{\sigma} = \breve{\tilde{\sigma}}$ (3),

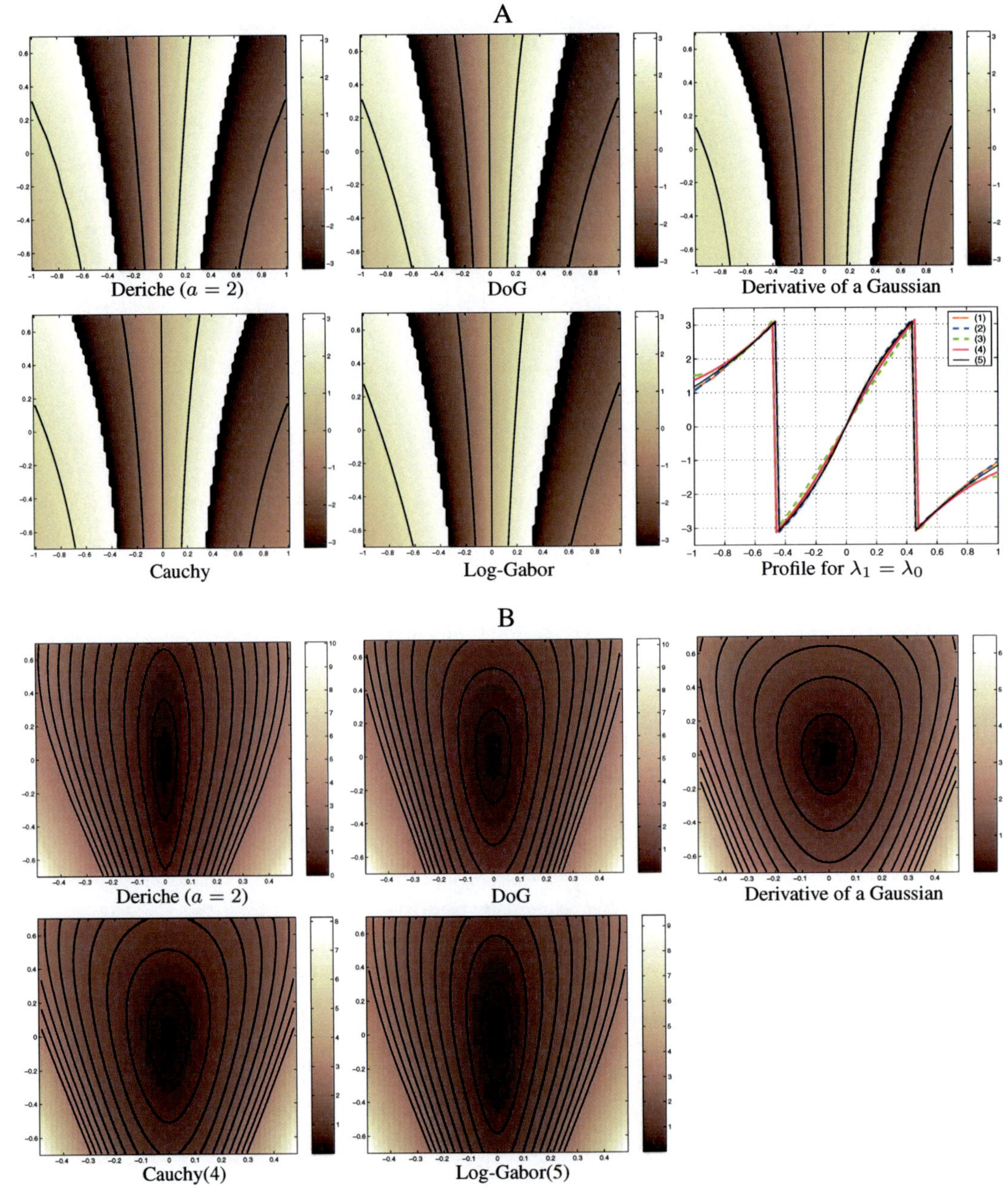

Figure 12: Mean phase difference level contours $\mu(z) = n\,\pi/2$, with $n = -4, \ldots, 4$ (A) and level contours of $b(z) = 0.3\,n$, with $n = 1, \ldots, 10$ (B), near $\mathbf{p}_0 = (0, \lambda_0)$ with a log scale on the vertical axis over two octaves $\frac{\lambda_1}{\lambda_0} \in [\frac{1}{2}, 2]$, and spatial position on the horizontal axis, $\frac{x_1}{\lambda_0} \in [-1, 1]$. All the filters have a bandwidth of 2.543 octaves.

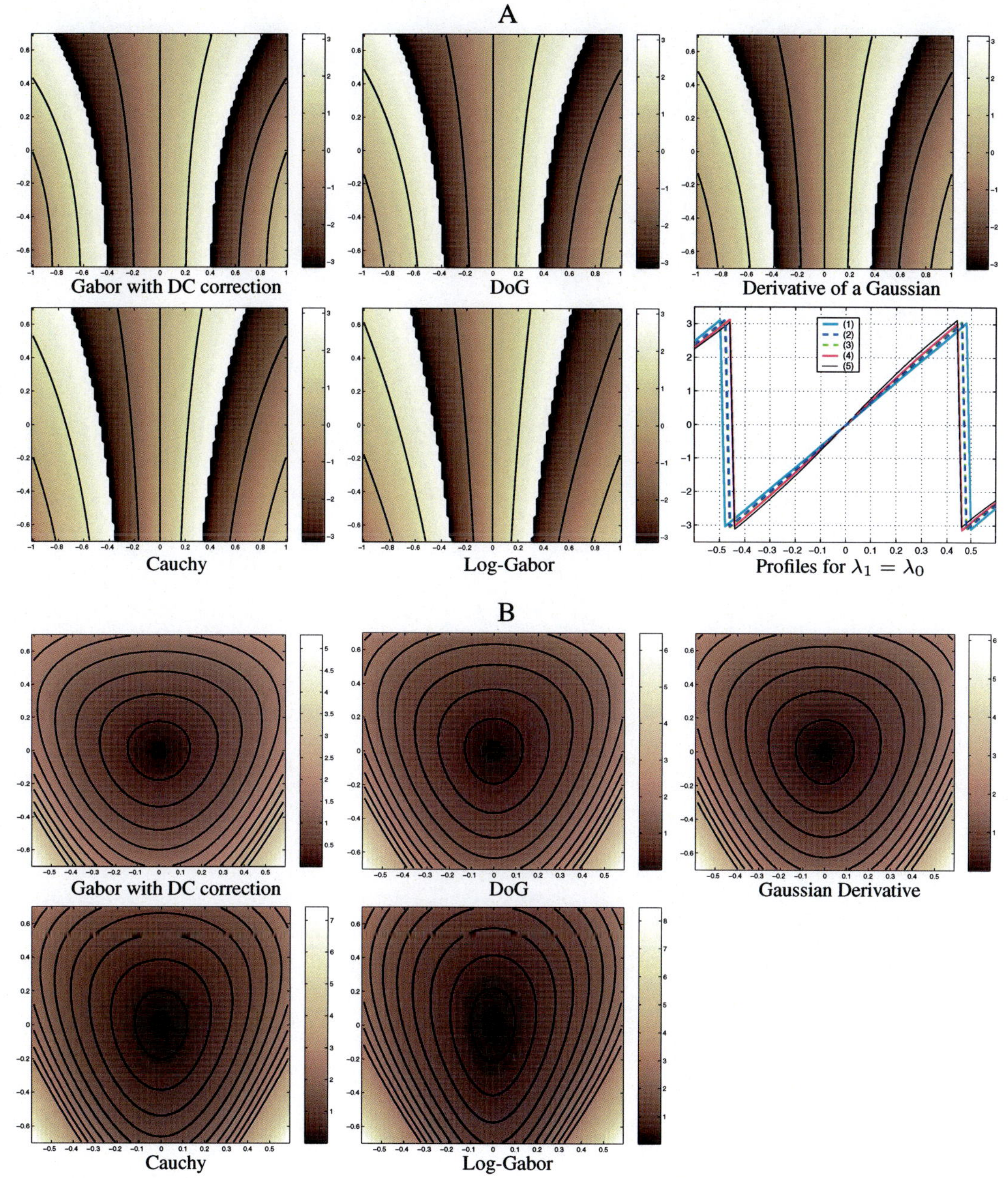

Figure 13: Mean phase difference level contours $\mu(z) = n\,\pi/2$, with $n = -4, \ldots, 4$ (A), and Level contours $b(z) = 0.3\,n$, with $n = 1, \ldots, 10$ (B), near $\mathbf{p}_0 = (0, \lambda_0)$ with a log scale on the vertical axis over two octaves $\frac{\lambda_1}{\lambda_0} \in [\frac{1}{2}, 2]$, and spatial position on the horizontal axis, $\frac{x_1}{\lambda_0} \in [-1, 1]$. The Gabor zero-DC is obtained using the proposed correction in (Appendix C). All the filters have a bandwidth of $\beta = 1.8$.

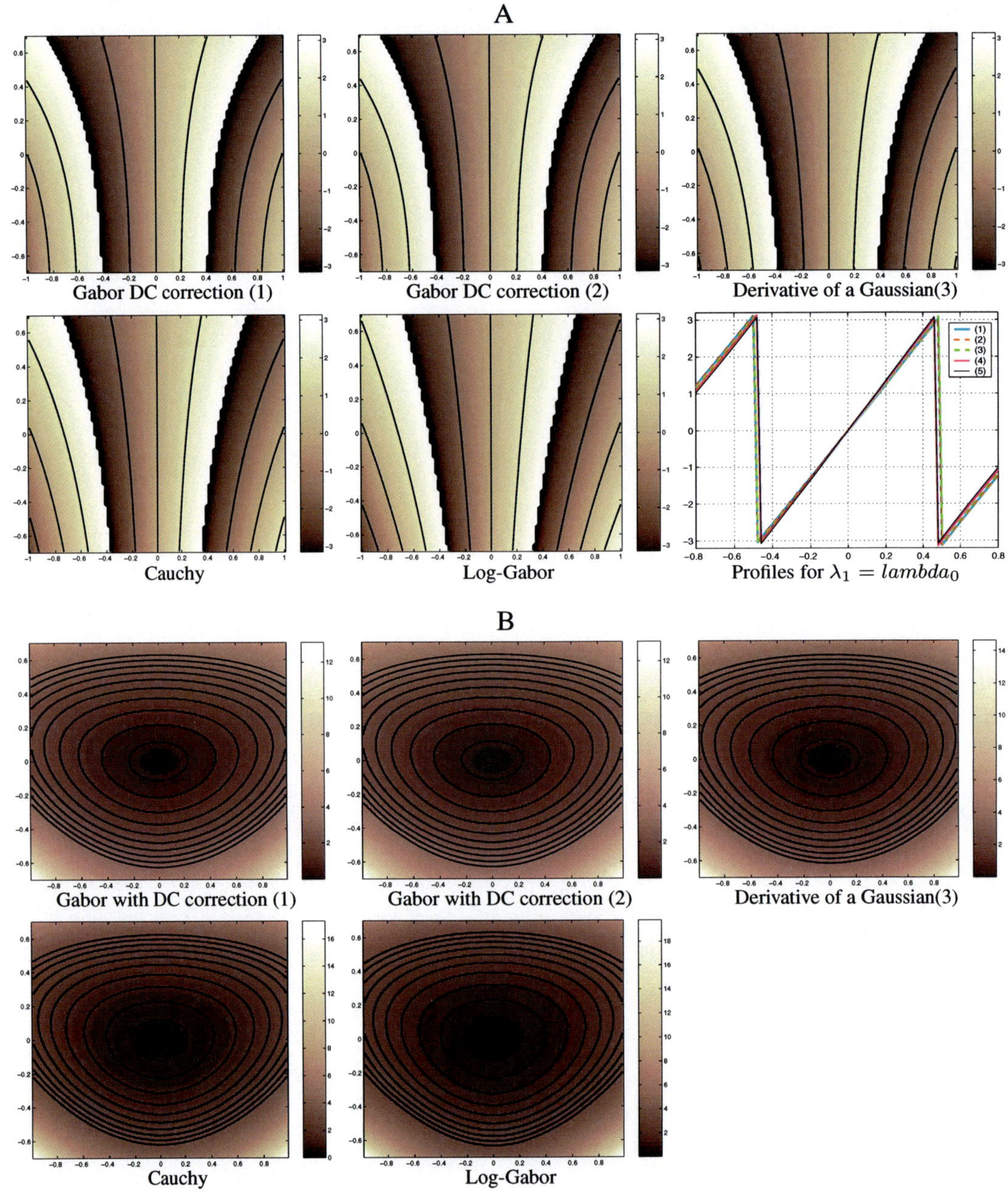

Figure 14: Mean phase difference level contours $\mu(z) = n\,\pi/2$, with $n = -4, \ldots, 4$ (A), and Level contours $b(z) = 0.3\,n$, with $n = 1, \ldots, 10$ (B), near $\mathbf{p}_0 = (0, \lambda_0)$ with a log scale on the vertical axis over two octaves $\frac{\lambda_1}{\lambda_0} \in [\frac{1}{2}, 2]$, and spatial position on the horizontal axis, $\frac{x_1}{\lambda_0} \in [-1, 1]$. Gabor (with (Appendix C) as zero-DC correction) (1); Gabor (with [78] as zero-DC correction) (2). All the filters have a bandwidth of $\beta = 1$.

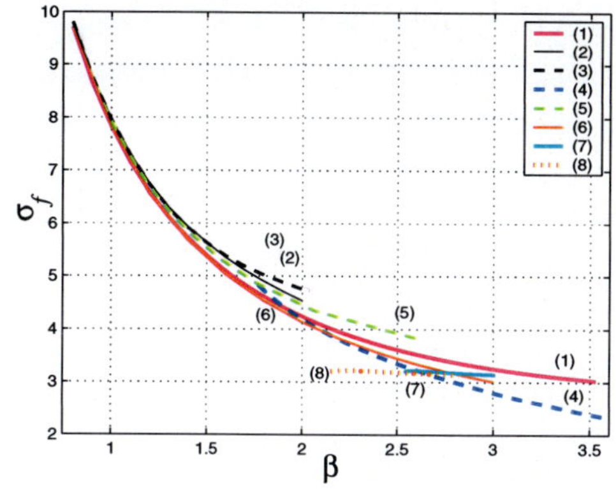

(a)

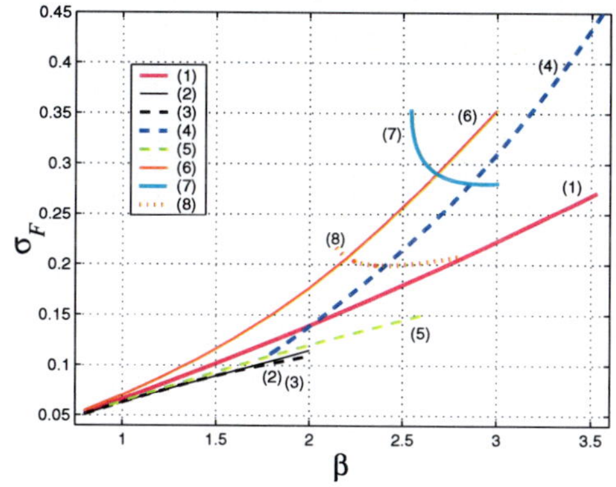

(b)

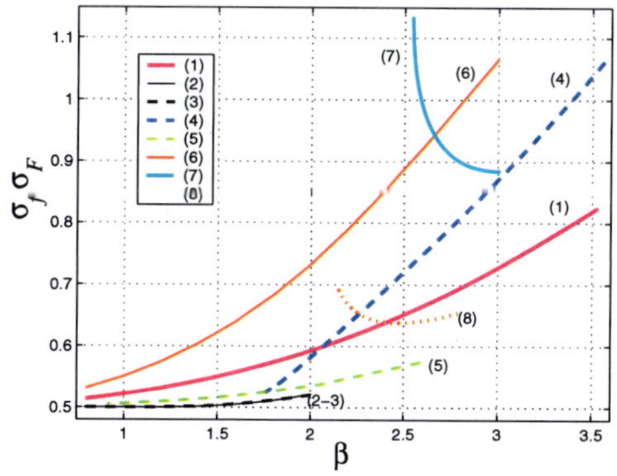

(c)

Figure 15: Uncertainty principle for the quadrature pairs for $\lambda_0 = 20$. Cauchy filters (curves (1)), Gabor with Ronse zero-DC correction using $\tilde{\sigma} = \sigma$ (curves (2)) and $\tilde{\sigma} = \check{\tilde{\sigma}}$ (curves (3)), DoG filters (curves (4)), Gaussian derivative filters (curves (5)), log-Gabor (curves (6)) and Sarkar and Boyer filters for best $\Gamma\Sigma$ (i.e Deriche filter) (curves (7)) and best $\Gamma\Sigma\Omega_S$ (curves (8)).

DoG filters (4), Gaussian derivative filters (5), log-Gabor (6) and Sarkar and Boyer filters for best $\Gamma\Sigma$ (i.e Deriche filter) (7) and best $\Gamma\Sigma\Omega_S$ (8).

Inspecting Fig. 15(a), which shows the spatial standard deviations of the filter's energy as a function of the bandwidth, we notice that all the families have a similar spatial extent for small bandwidths ($\beta < 1.5$). For larger bandwidths, the Gaussian derivative and the corrected Gabor wavelets have the largest σ_f. The three following families: Cauchy, log-Gabor and the DoG filters have a very similar σ_f (the DoG filter being the well localised when $\beta > 2.1$ and the worst for $1.76 < \beta < 1.9$). Interestingly, the dependence of σ_f on the bandwidth for the Deriche filter and the Sarkar and Boyer filter is very low.

In the Fourier domain, the corrected Gabor filters have the best localisation followed by the Gaussian derivative family (see Fig. 15(b)). The behaviour of the Cauchy filters is between the Gaussian derivatives and the log-Gabor, which has the worst frequency localisation. The DoG filter has a better localisation than Cauchy kernels for $1.76 < \beta < 2$ and is worse when $\beta > 2$. Here too, we notice that the dependence of σ_F on the bandwidth for Deriche filters and Sarkar and Boyer family is opposite to the dependence of other families. This result is expected as, for example, we already pointed out that the Deriche family has more aliasing for smaller bandwidths. This is because the bandwidth is inversely proportional to the order a of the derivative of the smoothing function and the decay of its Fourier transform is of order $\omega^{(a-4)}$. Notice that the Sarkar and Boyer filter has a far better frequency localisation than the Deriche one, especially for smaller bandwidths.

The relative behaviour of the different families regarding the uncertainty principle shown in Fig. 15(c) is similar to their relative behaviour regarding σ_F. The zero-DC corrected Gabor filter is close to the limit ($\frac{1}{2}$) almost everywhere, followed by the Gaussian derivatives filter. Because of their tailed nature, the log-Gabor wavelets are the farthest from the limit with the Deriche filter for $2.54 < \beta < 2.66$ (i.e $a \in [1.46, 2]$); the Sarkar and Boyer filters seems to have the best scale-space localisation for $2.590 < \beta \leq 2.787$ (basically for a bandwidth greater than the 1st derivative of a Gaussian until the bandwidth of Sarkar and Boyer's edge operator). For very large bandwidths, Cauchy kernels are the closest to the limit.

In the case of feature detection we can locate Canny's, Deriche and Sarkar and Boyer's edge operators as well as Ziou's line detector on these curves (Fig. 15, curves (5), (7), (8) and (7) respectively); these operators are optimal in the sense of Canny's criteria. This gives some insight on the performance of the other families, in the sense of Canny. Recently, the log-Gabor filters have been proposed as the choice of quadrature pair for feature detection. Spatial localisation of the filters is of great interest. On inspection of Fig. 15(a), we notice that the DoG filter has better spatial localisation than log-Gabors for a bandwidth greater than 2.1 octave and the Sarkar and Boyer filters are far better in their range of definition. When $\beta < 2$ log-Gabors are better than the DoG but the Cauchy filter has a comparative localisation. Fig. 15(b) shows that the DoG and the Cauchy family have always a better frequency localisation than log-Gabor kernels and Fig. 15(c) shows that the two families have a better uncertainty principle than the log-Gabor family. All three families are invariant to a grey level shift and to an additive ramp. Finally, note that the Deriche edge operator is closer to the DoG filter, and the Sarkar and Boyer edge operator is closer to the Cauchy family. Hence, for edge detection log-Gabor kernels are not the best choice.

However, for line detection, log-Gabor filters can have good performance as they are the closest to Ziou's optimal line detector. However, we found that the DoG filter for larger bandwidths has similar performance to the Ziou line detector. In addition Ziou's line detector has poor spectral localisation and the largest aliasing.

Our results suggests that the DoG filter is the best to use for combined line and edge detection. The Cauchy family is an interesting alternative, and we have the analytical expressions of the filters in the spatial and the frequency domain. Further experiments are needed to investigate their use in more detail.

4 Conclusion

We began by stressing the importance of phase information in computer vision, then provided the reader with a short summary of the definition of the local phase in one dimension, and its extension to multi-dimensional signals. In practice, the estimation of the local phase and the related quantity, local energy, uses a pair of band-pass quadrature filters. Several families of quadrature pairs have been proposed and applied in the literature, in a number of different applications. Most authors have not provided a reasonable justification for the use of a particular family apart from simplicity of use or the satisfaction of the zero-DC condition. This is particularly the case in feature detection where, for example, several authors have not considered the invariance of the filters to an additive ramp (smooth gradient). This property is desirable and has been noted in classical feature detection (see e.g. [8]). Our analysis attempts to put the justification on a sound foundation.

Specifically, in the chapter, we have studied mathematically the most common quadrature filters in the literature. We have provided either a mathematical expression or a numerical estimation for most filter properties. Key results and properties are summarised in Table 3.

In the case of feature detection, we presented the first attempt to use classical edge/line operators for the design of a quadrature pair. Although the main advantage of phase-based feature detection is that it is a unified detection approach (not only for edges or lines), we can learn a lot from the separate behaviour of the even part (symmetric detector, lines for example) and the odd part (asymmetric detector, edges for example). Interestingly, we found that most of the classical optimal operators, have large bandwidth and several authors point out that broad bandwidth filters have a better SNR in phase-based detection.

Our analysis shows that Gabor filters are not a good choice for feature detection. The odd part has poor performance in the sense of Canny in comparison to the 1st derivative of a Gaussian, to Sarkar and Boyer filter and to the Deriche filter; and the even part does not satisfy the zero-DC correction and hence needs correction. The zero-DC correction is possible but the filters are still not truly in quadrature, except if the negative frequencies are zeroed in the case of a Fourier domain implementation. Gabor filters are sensitive to smooth gradients.

We also concluded that log-Gabor kernels are not the best choice for feature detection. As we showed, the DoG family and the Cauchy family have better properties. The Sarkar and Boyer's filter has very good properties, for example for a bandwidth close to 2 octaves, this family has very good spatial localisation. In the case that one is interested only in one type of features (edges or lines), a natural choice is the Sarkar and Boyer family which is optimal in the sense of Canny.

Regarding the use of quadrature filters for velocity estimation, we showed that all the families have similar behaviours for small bandwidths (in terms of phase linearity and phase stability). The zero-DC corrected Gabor filters could be used. If one wanted the filter to be invariant to a smooth gradient (in the case where the conservation criteria is violated), the Gaussian derivative family would be a natural choice as they have close behaviour to the Gabor filters in terms of phase stability and the uncertainty principle. The disadvantage of using the Gaussian derivative filter is the loss of the spatial representation of the filters. In this case, the Cauchy family offers an alternative and has the advantage of being invariant to smooth gradients and having analytical expression in the spatial and the Fourier domain.

A Phase Stability Given White Gaussian Noise

The stability of phase can be measured in terms of the differences in phase response between two arbitrary neighbouring scale-space locations, $\mathbf{p}_0 = (x_0, \lambda_0)$, $\mathbf{p}_1 = (x_1, \lambda_1)$. We denote the filter response to the input $f(x)$ at scale-space position $\mathbf{p}_j$ by:

$$g_j = h(x - x_j,\, \lambda_j) \circledast f(x) \ = < h(x_j - x,\, \lambda_j),\, f(x) > \ , \tag{A.1}$$

where $< f(x), g(x) >= \int_{-\infty}^{\infty} f(x)^\star\, g(x)\, dx$ is the inner product.
The phase differences, as a function of the relative scale-space position, $\mathbf{p}_1 - \mathbf{p}_0 = (\Delta x, \Delta\lambda)$ can be written as:

$$\Delta\phi = \arg[g_1] - \arg[g_0] \ . \tag{A.2}$$

Phase is stable when $\Delta\phi$ is constant with respect to changes in scale $\Delta\lambda$; and it is linear with respect to spatial position when $\Delta\phi$ is linear function of Δx.

To characterise the stability of phase behaviour through scale-space, Fleet and Jepson [29] examined the response of the filter kernel, $h(x, \lambda)$ and its phase behaviour to stationary, zero mean white Gaussian noise. The authors derived an approximation to the mean phase difference $\mathrm{E}[\Delta\phi]$, and an upper bound of the mean variation about the mean $\mathrm{E}[|\Delta\phi - \mathrm{E}[\Delta\phi]|]$. Given the above noise assumptions:

$$E[\Delta\phi] \approx \mu(z) = \arg[z] \tag{A.3}$$

$$\mathrm{E}[|\Delta\phi - \mathrm{E}[\Delta\phi]|] \le b(z) = \frac{\sqrt{1 - |z|^2}}{|z|} \ , \tag{A.4}$$

where the complex scalar z is a function of scale-space position and is given by:

$$z(\mathbf{p}_1 - \mathbf{p}_0) = z(\Delta x,\, \Delta\lambda) \ = < h(x_0 - x,\, \lambda_0),\, h(x_1 - x,\, \lambda_1) > \ , \tag{A.5}$$

where the kernel h has to satisfy the two following conditions:

1. Normalisation condition:

$$\|h(x,\, \lambda)\| \ = < h(x,\, \lambda), h(x,\, \lambda) >= 1 \ . \tag{A.6}$$

2. Translational invariance and self-similarity across scales (i.e their octave bandwidth remains constant, independent of the scale to which they are tuned):

$$h(x,\, \lambda) = \frac{1}{\sqrt{\lambda}}\, h(x/\lambda,\, 1) \ . \tag{A.7}$$

B Optimal Linear Edge Detector Criteria

Consider the problem of detecting an ideal feature (edge or line) $f_F(x)$, centred at $x = 0$ using an observation corrupted by an additive white Gaussian noise with standard deviation $\sigma = 1$. Canny [8, 9] was the first to formulate the detection problem for a linear edge detector $f_O(x)$ as an optimisation problem. As Canny pointed out, the key to the design of an effective edge operator is accurate evaluation of its performance. Canny defined three different criteria to be maximised [9]:

Good Detection: Given the white Gaussian noise assumption, the probability of false detection decreases with increasing the signal-to-noise ratio $\Sigma(f)$ given by:

$$\Sigma(f_O) = \frac{\mid f_F(x) \circledast f_O(x) \mid}{\sqrt{\int_{-\infty}^{\infty} f_O^2(x)dx}} = \frac{\mid \int_{-\infty}^{\infty} f_F(-x) f_O(x) dx \mid}{\sqrt{\int_{-\infty}^{\infty} f_O^2(x)dx}} \quad . \tag{B.1}$$

Good Localisation: This second criterion is defined as the variance of the maximum of the filter response about the correct location [10]. Canny used an approximation based on a Taylor's expansion of the filter output [11]:

$$\text{for edges} \quad \Lambda(f_O) = \frac{\mid \int_{-\infty}^{\infty} f_F'(-x) f_O'(x) dx \mid}{\sqrt{\int_{-\infty}^{\infty} {f_O'}^2 dx}} \quad . \tag{B.2}$$

$$\text{for lines} \quad \Lambda(f_O) = \frac{\mid \int_{-\infty}^{\infty} f_F(-x) f_O''(x) dx \mid}{\sqrt{\int_{-\infty}^{\infty} {f_O'}^2 dx}} \quad . \tag{B.3}$$

Low Number of Multiple Responses: This criterion was designed to minimise the number of spurious maxima within the operator spatial spread [9]:

$$\Omega_C(f_O) = 2\pi \sqrt{\frac{\int_{-W}^{W} {f_O'}^2 dx}{\int_{-W}^{W} {f_O''}^2 dx}} = kW \ , \tag{B.4}$$

where W is the operator width and k is a positive constant.

Sarkar and Boyer [80] point out that the above multiple response criteria must be modified to be applied to IIR filters. In other words we should provide an estimate of the operator width W. Sarkar and Boyer [80] defined the width of the filter as the normalised root mean square deviation of the function $f_O^2(x)$ from the origin.

$$\bar{W} = \sqrt{\frac{\int_{-\infty}^{\infty} x^2 f_O^2(x) dx}{\int_{-\infty}^{\infty} f_O^2(x) dx}} \quad . \tag{B.5}$$

---

[9] Canny considered the 1D continuous case; Canny-like criteria for discrete sampled signals are considered in [14]; 2D steerable filters using Canny-like criteria are proposed in [42].

[10] Although the agreement [6, 80] and the disagreement [86], this is the most popular localisation measure in this context.

[11] The Taylor expansion of the filters output are different for an edge detector and a line detector because the edge detector is odd and the line detector is even. See Canny [9] for the details.

Hence, they proposed a new expression for $\Omega(f)$:

$$\Omega_S(f_O) = 2\pi \sqrt{\frac{\int_{-\infty}^{\infty} f_O'^2 dx}{\int_{-\infty}^{\infty} f_O''^2 dx}} \sqrt{\frac{\int_{-\infty}^{\infty} f_O^2(x) dx}{\int_{-\infty}^{\infty} x^2 f_O^2(x) dx}} , \tag{B.6}$$

Then, the optimality of the detector is defined with respect to the maximisation of the following quantity [9]:

$$C_c(f_O) = \Gamma(f_O)\, \Sigma(f_O)\, \Omega(f_O) . \tag{B.7}$$

For an ideal step edge (B.7) is given by:

$$C_c(f_O) = \frac{\mid \int_{-\infty}^{0} f_O(x) dx \mid\mid f_O'(0) \mid}{\sqrt{\int_{-\infty}^{\infty} f_O^2(x) dx \int_{-\infty}^{\infty} f_O''^2 dx}} . \tag{B.8}$$

C Gabor DC Correction

We have seen that Gabor filters are not invariant to a grey level shift of the input because they do not have zero-DC. In this appendix, we study this problem and compare a number of zero-DC corrections that have been proposed in the literature. A variant to one of them, which is optimal in terms of the reduction of the power of the negative frequencies, with a correction to the tuning frequency shift, is also proposed.

C.1 Gabor DC Correction: Heitger et al.

In order the achieve the zero-DC constraint, Heitger et al [36, 37] proposed the following family:

$$g_h(x) = n_c \frac{1}{\sqrt{2\pi}} \exp\left(-\frac{x^2}{2\sigma^2}\right) e^{[i\,\omega_0\, x\, \psi(x)]} , \tag{C.1}$$

$$\text{where} \quad \psi(x) = \frac{\text{erf}(\tau\, x)}{x} \quad \text{and} \quad \tau \geq 0 . \tag{C.2}$$

Here $\psi(x)$ is the frequency sweep function and τ is chosen numerically to achieve zero-DC. Note first that Heitger filters are equivalent to Gabor filters when $\tau = 0$. Second, the sweep function, as given in eq.(C.2), is not well parameterised since the value of τ depends on σ and the tuning frequency will be different from ω_0 (this is because $\psi(0) \neq 1$). Taking into account these two observations, we propose the use of the following parameterisation:

$$\psi_1(x) = \frac{\sqrt{\pi}}{2\tau} \frac{\text{erf}(\tau \frac{x}{\sigma})}{\frac{x}{\sigma}} . \tag{C.3}$$

A second sweep function, with two degrees of freedom, was also proposed in [37]; it is given by the following:

$$\psi_2(x) = 1 - \tilde{k}\left(1 - \exp[-\tau \frac{x^2}{\sigma^2}]\right) \quad \text{with,} \quad 0 \leq \tilde{k} \leq 1 . \tag{C.4}$$

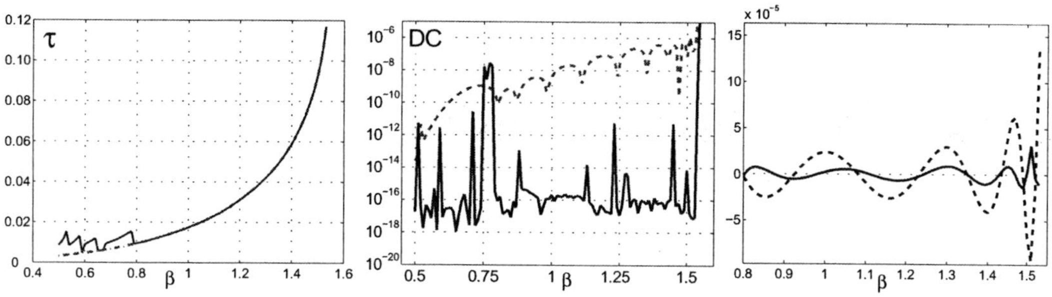

Figure C.1: Left: τ values for ψ_2 with $\tilde{k} = 0.5$; estimated numerically (solid lines); model predicted (dashed lines). Middle: DC values after correction using the previous τ values. Right: Relative fitting error between the numerically estimated τ values and the model predicted values for ψ_1 (dashed curve) and ψ_2 (solid curve).

For instance, ψ_1 was used for feature detection in [36], and ψ_2 in [46]. This family of functions are called S-Gabor filters (Stretched-Gabor).

We estimated numerically the values of τ and the DC value for both functions when $0.5 \leq \beta \leq 1.55$. The results are shown for ψ_2 in Fig. C.1. The DC values should be compared to those without correction shown in Fig. C.3. Note that the DC component for the numerically calculated values of τ is around 10^{-16} (i.e vanishingly small). We used the following parametric function to approximate the evolution of τ function of the bandwidth β: $\tau(\beta) = \sum_{j=1}^{6} a_j \exp(b_j\, \beta^j) + c_j\, \beta^{(j-1)}$. This function provides a good fit for the τ values as can be seen in Fig. C.1. The relative fitting error is less than 2×10^{-5} for ψ_2 and 10^{-4} for ψ_1. Unfortunately, inspecting Fig. C.1, we see that the DC values obtained using the fitting function are 10^8 greater than those that are estimated numerically; but they are still $\leq 10^{-6}$.

There are two main drawbacks with these families. First, there is not an analytical expression for their Fourier representation. Second, zero-DC correction exists only for $\beta < 1.55$ [46] (see the right side of Fig. C.1). A third (minor) problem is that the value of τ is not unique and is very sensitive (i.e. we need a high precision of estimation). The correct value is the smallest one, since when $\tau \to 0$ the filters are equivalent to Gabor filters. Thus we correct the DC component with a minimum distortion to the filter's shape.

This final point is important, as we have noticed that in a recent paper [87], the authors used the S-Gabor filters for contour completion using ψ_2 with $k = 0.5$ as a sweep function. They used a filter with $\sigma = 2.0$ tuned at a frequency corresponding to $\lambda_0 = 5$ ($\beta = 1.466$ octaves) and a value of $\tau = 0.334$. For this τ value, the filter has zero DC but it is not the smallest one as can be seen from Fig. C.1. The smallest value for this bandwidth is around 0.0776. The S-Gabor filters are plotted in Fig. C.2(a) for the two values of τ as well as the Gabor filters for the same tuning frequency and bandwidth. Notice the distortion of the filters in terms of the tuning frequency and the bandwidth and especially the increase of the power of the negative frequencies when using the wrong value of τ (0.334). This implies that the filters are not truly in quadrature, which may introduce errors in the estimation of local image information (phase, energy).

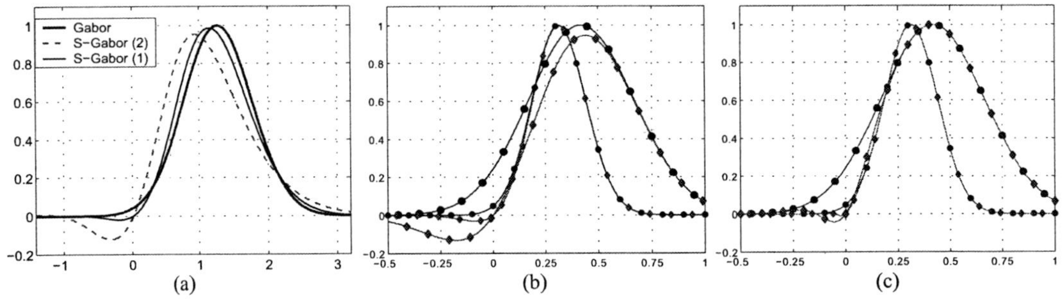

Figure C.2: (a) Gabor filters for $\sigma = 2.0$, $\lambda_0 = 5$ (thick curves). S-Gabor (1) using ψ_2 for $\tilde{k} = 0.5$, and $\tau = 0.0776$ (thin curves). S-Gabor (2) are obtained using the same parameters except $\tau = 0.334$ (dashed curves). (b) Corrected Gabor filter using [78] (diamond curves). Original Gabor filters (circle curves) with $\beta = 1.5$, $\lambda_0 = 20$ and $\beta = 2.5$, $\lambda_0 = 15$. (c) is the same as (b) using the proposed correction i.e for $\tilde{\sigma} = \check{\tilde{\sigma}}$.

C.2 Gabor DC Correction: Ronse

In order to correct the even Gabor filter, Ronse suggested subtracting a smooth function (a Gaussian) as follows:

$$g_r(x) = n_c \frac{e^{-\frac{x^2}{2\sigma^2}}}{\sqrt{2\pi}\,\sigma} \left(e^{i\,\omega_0\,x} - e^{-\frac{\sigma^2\,\omega_0^2}{2}} \right) , \tag{C.5}$$

$$G_r(\omega) = n_c\, e^{-\frac{\sigma^2}{2}(\omega-\omega_0)^2} - n_c\, e^{-\frac{\sigma^2}{2}(\omega^2+\omega_0^2)} . \tag{C.6}$$

We observe that the filter has zero-DC, however, this correction does not solve the problem of the filter not being in quadrature. (The filter still has a considerable amount of power in the negative frequencies).

Figure C.2 (b) shows the DC-corrected filter in the frequency domain as well as the uncorrected Gabor filter. We note that Ronse's DC correction introduces some undesirable artifacts. First, note that there is a small shift of the tuning frequency towards the high frequencies and hence introduces a small reduction in the desirable bandwidth. A second problem with this correction is that the negative power of the filter increases as demonstrated in Fig. C.3.

Equations (C.5) and (C.6) can be written in the following form with $\tilde{\sigma} = \sigma$:

$$\tilde{g}_r(x) = n_c \left(\frac{e^{-\frac{x^2}{2\sigma^2}}}{\sqrt{2\,\pi\sigma}}\, e^{i\,\omega_0\,x} - \frac{e^{-\frac{x^2}{2\tilde{\sigma}^2}}}{\sqrt{2\,\pi\tilde{\sigma}}}\, e^{-\frac{\sigma^2\,\omega_0^2}{2}} \right) , \tag{C.7}$$

$$\tilde{G}_r(\omega) = n_c\, e^{-\frac{\sigma^2}{2}(\omega-\omega_0)^2} - n_c\, e^{-\frac{1}{2}(\tilde{\sigma}^2\,\omega^2+\sigma^2\,\omega_0^2)} . \tag{C.8}$$

It is easy to show that the zero-DC condition is satisfied for all $\tilde{\sigma} > 0$. It is possible to find an optimal value for $\tilde{\sigma}$ that reduces the power of the negative frequencies. Hence the optimal value for $\tilde{\sigma}$ in this sense is defined by:

$$\hat{\tilde{\sigma}} = \arg\min_{\tilde{\sigma}} \int_{-\infty}^{0} \tilde{G}_r(\omega)^2\, d\omega \qquad \text{with,} \tag{C.9}$$

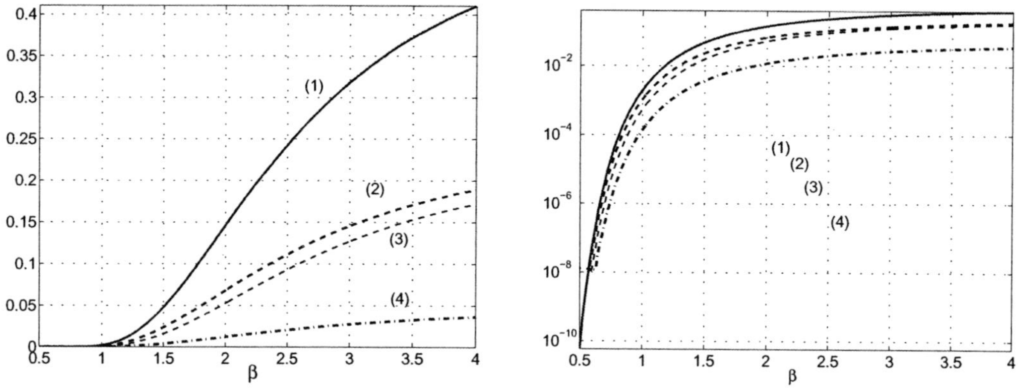

Figure C.3: DC_{g_b} value for Gabor filters (without correction)(1). Power of the negative frequencies (without correction)(3); with correction for: $\tilde{\sigma} = \check{\tilde{\sigma}}$ (4), $\tilde{\sigma} = \sigma$ (2) i.e using [78]. Left, linear scale; Right, logarithmic scale.

$$\int_{-\infty}^{0} \tilde{G}_r(\omega)^2\, d\omega = \frac{n_c^2 \sqrt{\pi}}{2} \left\{ \frac{1 - \mathrm{erf}[\sigma\omega_0]}{\sigma} + \frac{\exp(-\sigma^2\, \omega_0^2)}{\tilde{\sigma}} \right.$$
$$\left. - \frac{2\sqrt{2}\left(1 - \mathrm{erf}[\frac{\sigma^2\, \omega_0}{\sqrt{2\,(\tilde{\sigma}^2+\sigma^2)}}]\right) \exp\left(-\frac{2\,\tilde{\sigma}^2+\sigma^2}{2\,(\tilde{\sigma}^2+\sigma^2)}\,\sigma^2\,\omega_0^2\right)}{\sqrt{\tilde{\sigma}^2+\sigma^2}} \right\}.$$

Finding an analytical solution to the minimisation problem given in (C.9) is difficult. As an alternative, we minimised the integral of the negative frequencies rather than its energy:

$$\int_{-\infty}^{0} \tilde{G}_r(\omega)\, d\omega = n_c \sqrt{\frac{\pi}{2}} \left(\frac{1 - \mathrm{erf}[\frac{\sigma\omega_0}{\sqrt{2}}]}{\sigma} - \frac{e^{-\sigma^2\,\omega_0^2/2}}{\tilde{\sigma}} \right).$$

It can be shown that the above integral is equal to zero for:

$$\check{\tilde{\sigma}} = \frac{\exp[-\sigma^2\,\omega_0^2/2]}{1 - \mathrm{erf}[\sigma\omega_0/\sqrt{2}]}\,\sigma\ . \tag{C.10}$$

We calculated the optimal $\hat{\tilde{\sigma}}$ given by (C.9) numerically and its corresponding negative energy as well as for the approximated value $\check{\tilde{\sigma}}$ given by (C.10). We found that the negative energy for $\check{\tilde{\sigma}}$ is always close to the global minimum of (C.9) (the relative error is always less than 4%). We conclude that $\check{\tilde{\sigma}}$ is an acceptable approximation for $\hat{\tilde{\sigma}}$.

Figure C.2 (c) is the corresponding figure to Fig. C.2 (b) for $\tilde{\sigma} = \check{\tilde{\sigma}}$. Notice that the shift of the tuning frequency and the power of the negative frequencies are both considerably reduced. Figure C.3 shows the power of the negative frequencies function of the bandwidth of the corrected Gabor filter for $\tilde{\sigma} = \check{\tilde{\sigma}}$; the equivalent curves of the uncorrected Gabor filters and of the corrected for $\tilde{\sigma} = \check{\tilde{\sigma}}$ are also shown for comparison. Using the optimal value $\check{\tilde{\sigma}}$ reduces the power of the negative frequencies by a factor of 5.

Next, we study the influence of the DC correction on the tuning frequency and on the bandwidth of the filters. To do this we have to find the argument of the maximum of the corrected filters in the frequency domain (eq. C.8). In essence, as we know that there is only

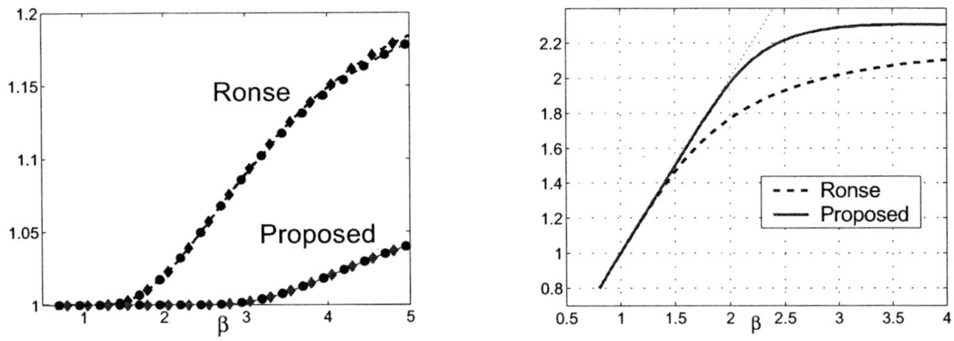

Figure C.4: Influence of the DC correction on the tuning frequency and on the bandwidth. DC correction with $\tilde{\sigma} = \sigma$ (dashed curves) and optimal $\tilde{\sigma} = \check{\tilde{\sigma}}$ (solid curves). Left: Ratio of the tuning frequency after correction $\hat{\omega}_0$ on the tuning frequency before correction ω_0: numerically estimated (diamond curves), approximated using (C.12) (cirles curves). Right: True bandwidth after correction function of the bandwidth before correction.

one maxima, we can try to find the zero-crossing of the first derivative of the Fourier transform of the filters. Unfortunately, this leads to a non-linear equation. Hence as we know that the DC correction introduces a small shift of the tuning frequency, we can approximate the first derivative of the filters around ω_0 using its Taylor expansion. Setting $\omega = \omega_0 + d\omega$, we can show that the Taylor expansion of the first derivative of eq. (C.8) around ω_0 is given by:

$$\begin{aligned} & a_0 + a_1\, d\omega + a_2\, d\omega^2 + \mathrm{O}(d\omega^3) \ , \qquad \text{(C.11)} \\ \text{with,} \quad & a_0 = \omega_0\, \tilde{\sigma}^2 \exp[-\omega_0^2\, (\sigma^2 + \tilde{\sigma}^2)/2] \ , \\ & a_1 = (\tilde{\sigma}^2 - \omega_0^2\, \tilde{\sigma}^4) \exp[-\omega_0^2\, (\sigma^2 + \tilde{\sigma}^2)/2] - \sigma^2 \ , \\ & a_2 = \frac{\omega_0 \tilde{\sigma}^4}{2} (\omega_0^2\, \tilde{\sigma}^2 - 3) \exp[-\omega_0^2\, (\sigma^2 + \tilde{\sigma}^2)/2] \ . \end{aligned}$$

Hence, the true tuning frequency can be approximated by:

$$\hat{\omega}_0 = \omega_0 + \hat{d\omega} \qquad \text{(C.12)}$$

where $\hat{d\omega}$ is the positive solution of eq. (C.11) set to zero.

Figure C.4 shows that $\hat{\omega}_0$ is a good approximation to the true tuning frequency. Notice, that the frequency shift and the bandwidth reduction are considerably reduced for $\tilde{\sigma} = \check{\tilde{\sigma}}$ than when $\tilde{\sigma} = \sigma$ and that the bandwidth of the corrected filters is upper bound by 2.4 octaves.

C.3 Comparison of the DC Corrections

The advantage of the Ronse [78] correction is that there is an analytical expression for the Fourier transform of the filters and larger bandwidth filters ($\beta < 2.4$) can be defined. This is not the case for the Heitger et al. [37] correction.

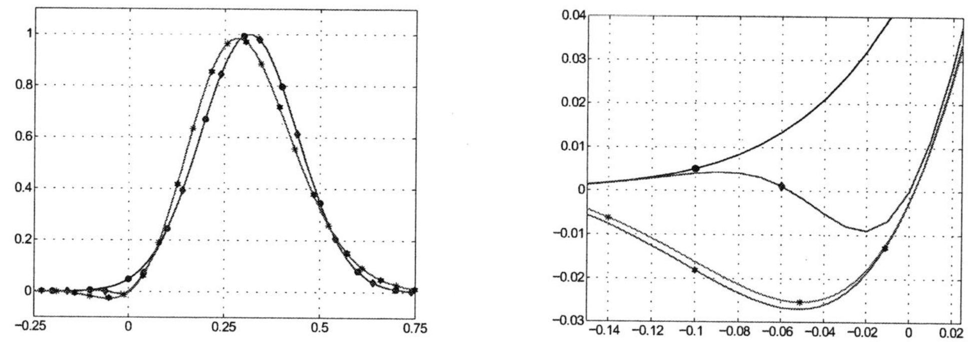

Figure C.5: Comparison of the DC corrections. Gabor (circles curve), correction with [78] for $\tilde{\sigma} = \check{\tilde{\sigma}}$ (diamond curve); correction with [37] using ψ_1 (asterix curve) and using ψ_2 (stars curve). Zoom at the zero frequency (right).

Figure C.5 shows both corrections in the frequency domain. Note that the power of the negative frequencies using the Heitger correction is greater than using the Ronse correction with $\check{\tilde{\sigma}}$. The zoom on the zero frequency (Fig. C.5, right) shows that the DC component is small but still not equal to zero for the Heitger correction. A second problem with Heitger filters is the large shift of the tuning frequency towards the low frequencies, and therefore a modification of the true bandwidth. Not having an analytical expression of the Fourier transform of the filters discards any analytical correction of this shift.

Acknowledgements

This work was partially funded by the EC project ADEQUATE (IST: 1999-10837) and the UK EPSRC funded IRC in Medical Image Analysis and Signals (MIAS).

References

[1] Aach, T., Kaup, A. and Mester, R.: 1995, On texture analysis: Local energy transforms versus quadrature filters, *Signal Processing* **45**(2), 173–181.

[2] Barron, J. L., Fleet, D. J. and Beauchemin, S. S.: 1994, Performance of optical flow techniques, *International Journal of Computer Vision* **12**(1), 43–77.

[3] Bharath, A. A.: 1998, Steerable filters from Erlang functions, *British Machine Vision Conference*.

[4] Boukerroui, D., Noble, J. A. and Brady, M.: 2004, On the choice of band-pass quadrature filters, *Journal of Mathematical Imaging and Vision* **21**(1), 53–80.

[5] Bovik, A. C., Clark, M. and Geisler, W. S.: 1990, Multichannel texture analysis using localized spatial filters, *IEEE Trans. Pattern Anal. Machine Intell.* **12**(1), 55–73.

[6] Boyer, K. L. and Sarkar, S.: 1994, Comments on "On the localization performance measure and optimal edge detection", *IEEE Trans. Pattern Anal. Machine Intell.* **16**(1), 106–108.

[7] Bülow, T. and Sommer, G.: 2001, Hypercomplex signals - A novel extension of the analytic signal to the multidimensional case, *IEEE Trans Signal Processing* **49**(11), 2844–2852.

[8] Canny, J.: 1983, Finding edges and lines in images, *Technical Report Memo 720*, MIT Artif. Intell. Lab., Massachusetts Inst. Technol.

[9] Canny, J.: 1986, A computational approach to edge detection, *IEEE Trans. Pattern Anal. Machine Intell.* **8**(6), 679–698.

[10] Casasent, D. and Smokelin, J. S.: 1994, Real, imaginary, and clutter Gabor filter fusion for detection with reduced false alarms, *Optical Engineering* **33**, 2255–2263.

[11] Chen, C.-M., Lu, H. H.-S. and Han., K.-C.: 2001, A textural approach based on Gabor functions for texture edge detection in ultrasound images, *Ultrasound in Medicine & Biology* **27**(4), 515–534.

[12] Corless, R. M., Gonnet, G. H., Hare, D. E. G., Jeffrey, D. E. and Knuth, D. E.: 1996, On the Lambert W function, *Advances in Computational Mathematics* **5**, 329–359. **URL:** *http://pineapple.apmaths.uwo.ca/~rmc/papers/LambertW/index.html*

[13] Daugman, J. G.: 19, Two-dimensional spectral analysis of cortical receptive field profiles, *Vision Research* **10**, 847–856.

[14] Demigny, D.: 2002, On optimal linear filtering for edge detection, *IEEE Trans Image Processing* **11**(7), 728–737.

[15] Deriche, R.: , Recursively implementing the gaussian and its derivatives, *Technical Report 1893*, 1993, INRIA, Unité de Recherche Sophia-Antipolis.

[16] Deriche, R.: 1987, Using Canny's criteria to derive a recursively implemented optimal edge detector, *International Journal of Computer Vision* **1**(2), 167–187.

[17] Deriche, R.: 1990, Fast algorithms for low-level vision, *IEEE Trans. Pattern Anal. Machine Intell.* **12**(1), 78–87.

[18] Dunn, D. F. and Higgins, W. E.: 1995, Optimal Gabor filters for texture segmentation, *IEEE Trans Image Processing* **4**(7), 947–964.

[19] Elad, M., Teo, P. and Hel-Or, Y.: 1999, Optimal filters for gradient-based motion estimation, *International Conference on Computer Vision*, Corfu, Greece, pp. 559–565.

[20] Elder, J. H. and Zucker, S. W.: 1998, Local scale control for edge detection and blur estimation, *IEEE Trans. Pattern Anal. Machine Intell.* **20**(7), 699–716.

[21] Felsberg, M.: 2002, Disparity from monogenic phase, *in* L. v. Gool (ed.), *24. DAGM Symposium Mustererkennung, Zürich*, Vol. 2449 of *Lecture Notes in Computer Science*, Springer, Heidelberg, pp. 248–256.

[22] Felsberg, M. and Sommer, G.: 2000a, The multidimensional isotropic generalization of quadrature filters in geometric algebra, *in* S. G. and Y. Y. Zeevi (eds), *Proc. Int. Workshop on Algebraic Frames for the Perception-Action Cycle*, Vol. 1888 of *Lecture Notes in Computer Science*, Springer-Verlag, Heidelberg, Kiel, Germany, pp. 175–185.

[23] Felsberg, M. and Sommer, G.: 2000b, A new extension of linear signal processing for estimating local properties and detecting features, *in* G. Sommer, N. Krüger and C. Perwass (eds), *22. DAGM Symposium Mustererkennung, Kiel*, Springer-Verlag, Heidelberg, pp. 195–202.

[24] Felsberg, M. and Sommer, G.: 2001a, The monogenic signal, *IEEE Trans Signal Processing* **49**(12), 3136–3144.

[25] Felsberg, M. and Sommer, G.: 2001b, Scale adaptive filtering derived from the Laplace equation, *in* B. Radig and S. Florczyk (eds), *23. DAGM Symposium Mustererkennung, München*, Vol. 2191 of *Lecture Notes in Computer Science*, Springer-Verlag, Heidelberg, pp. 124–131.

[26] Felsberg, M. and Sommer, G.: 2004, The monogenic scale-space: A unifying approach to phase-based image processing in scale-space, *Journal of Mathematical Imaging and Vision* **21**, 5–26.

[27] Field, D. J.: 1987, Relations between the statistics of natural images and the response properties of cortical cells, *Journal of the Optical Society of America* **A**(4), 2379–2394.

[28] Fleet, D. J. and Jepson, A. D.: 1990, Computation of component image velocity from local phase information, *International Journal of Computer Vision* **5**(1), 77–104.

[29] Fleet, D. J. and Jepson, A. D.: 1993, Stability of phase information, *IEEE Trans. Pattern Anal. Machine Intell.* **15**(12), 1253–1268.

[30] Freeman, W. T. and Adelson, E. H.: 1991, The design and use of steerable filters, *IEEE Trans. Pattern Anal. Machine Intell.* **13**(9), 891–906.

[31] Gabor, D.: 1946, Theory of communication, *J. IEE (London)* **93**(III), 429–457.

[32] Granlund, G. H. and Knutsson, H.: 1995, *Signal Processing for Computer Vision*, Kluwer.

[33] Grigorescu, S. E., Petkov, N. and Kruizinga, P.: 2002, Comparison of texture features based on Gabor filters, *IEEE Trans Image Processing* **11**(10), 1160–1167.

[34] Hahn, S.: 1996, *Hilbert transforms in signal processing*, Artech House, Boston.

[35] Hancock, E.: 1993, Resolving edge-line ambiguities using probabilistic relaxation, *IEEE Computer Vision and Pattern Recognition*, pp. 300–306.

[36] Heitger, F.: 1995, Feature detection using suppression and enhancement, *Technical Report NR.163*, Image Science Lab, ETH-Zurich, Switzerland.

[37] Heitger, F., Rosenthaler, L., von der Heydt, R., Peterhans, E. and Kübler, O.: 1992, Simulation of neural contour mechanisms: From simple to the end-stopped cells, *Vision Research* **32**, 963–981.

[38] Hemmendorff, M., Andersson, M., Kronander, T. and Knutsson, H.: 2002, Phase-based multidimensional volume registration, *IEEE Trans Med. Imag.* **21**(12), 1536–1543.

[39] Highnam, R. and Brady, M.: 1999, *Mammographic Image Analysis*, Kluwer.

[40] Huang, C. L. and Chen, Y. T.: 1995, Motion estimation method using a 3D steerable filter, *Image & Vision Computing* **13**(1), 21–32.

[41] Idrissa, M. and Acheroy, M.: 2002, Texture classification using Gabor filters, *Pattern Recognition Letters* **23**(9), 1095–1102.

[42] Jacob, M. and Unser, M.: 2003, Design of steerable filters for feature detection using Canny-like criteria, *IEEE Trans. Pattern Anal. Machine Intell.* **26**(9), 1007–1019.

[43] Jain, A. and Farrokhnia, F.: 1991, Unsupervised texture segmentation using Gabor filters, *Pattern Recognition* **24**(12), 1167–1186.

[44] Jepson, A. and Jenkin, M. R. M.: 1989, The fast computation of disparity from phase differences, *IEEE Computer Vision and Pattern Recognition*, San Diego, CA, USA, pp. 398–403.

[45] Kennedy, L. M. and Basu, M.: 1999, A gaussian derivative operator for authentic edge detection and accurate edge localization, *International Journal of Pattern Recognition and Artificial Intelligence* **13**(3), 367–380.

[46] Klein, A., Lee, F. and Amini, A.: 1997, Quantitative coronary angiography with deformable spline models, *IEEE Trans Med. Imag.* **16**(5), 468–482.

[47] Klein, S. A. and Levi, D. M.: 1985, Hyperacuity thresholds of 1 sec: theoretical predictions and empirical validation, *Journal of the Optical Society of America* **2**(7), 1170–1190.

[48] Knutsson, H. and Andersson, M.: 2003, What's so good about quadrature filters?, *IEEE International Conference on Image Processing*, Barcelona. Invited Paper.

[49] Kovesi, P.: 1999, Image features from phase congruency, *Videre A Journal of Computer Vision Research* **1**(3), 1–26.

[50] Kruizinga, P. and Petkov, N.: 1999a, Nonlinear operator for blob texture segmentation, *IEEE Workshop on Nonlinear Signal Processing*, Antalya, Turkey, pp. 881–885.

[51] Kruizinga, P. and Petkov, N.: 1999b, Nonlinear operator for oriented texture, *IEEE Trans Image Processing* **8**(10), 1395–1407.

[52] Kruizinga, P. and Petkov, N.: 2000, Computational model of dot-pattern selective cells, *Biological Cybernetics* **83**(4), 313–325.

[53] Kuglin, C. D. and Hines, D. C.: 1975, The phase correlation image alignment method, *Int. Conf. Cybernetics and Society*, pp. 163–165.

[54] Larkin, K. G., Bone, D. J. and Oldfield, M. A.: 2001, Natural demodulation of two-dimensional fringe patterns: I. general background of the spiral phase quadrature transform, *Journal of the Optical Society of America* **18**(8), 1862–1870.

[55] Lee, T. S.: 1996, Image representation using 2D Gabor wavelets, *IEEE Trans. Pattern Anal. Machine Intell.* **18**(10), 959–971.

[56] Leite, J. A. F. and Hancock, E. R.: 1994, Statistically combining and refining multi-channel information, *in* S.Impedovo (ed.), *progress in image analysis and processing III*, World Scientific, pp. 193–200.

[57] Little, J.: 1992, Accurate early detection of discontinuities, *Vision Interface* pp. 97–102.

[58] Liu, H. C., Hong, T. H., Herman, M. and Chellappa, R.: 1997, A general motion model and spatiotemporal filters for computing optical-flow, *International Journal of Computer Vision* **22**(2), 141–172.

[59] Lowe, D. G.: 1999, Object recognition from local scale-invariant features, *International Conference on Computer Vision*, Vol. II, pp. 1150–1157.

[60] Mallat, S. G.: 1999, *A Wavelet Tour of Signal Processing*, Academic Press.

[61] Marčelja, S.: 1980, Mathematical description of the responses of simple cortical cells, *Journal of the Optical Society of America* **10**, 1297–1300.

[62] Marr, D. and Hildreth, E.: 1980, Theory of edge detection, *Royal Society London* **B-207**, 187–217.

[63] Marroquin, J., Figueroa, J. and Servin, M.: 1997, Robust quadrature filters, *Journal of the Optical Society of America* **14**(4), 779–791.

[64] Mellor, M. and Brady, M.: 2004, Non-rigid multimodal image registration using local phase, *MICCAI*, Vol. I of *Lect. Note Comput. Sci*, Renne-Sait-Malo, France, pp. 763–770.

[65] Mikolajczyk, K. and Schmid, C.: 2001, Indexing based on scale invariant interest points, *International Conference on Computer Vision*, pp. I: 525–531.

[66] Morrone, M. C., Navangione, A. and Burr, D.: 1995, An adaptive approach to scale selection for line and edge-detection, *Pattern Recognition Letters* **16**(7), 667–677.

[67] Morrone, M. C. and Owens, R. A.: 1987, Feature detection from local energy, *Pattern Recogn. Lett.* **6**(5), 303–313.

[68] Mulet-Parada, M. and Noble, J. A.: 2000, 2D+T acoustic boundary detection in echocardiography, *Medical Image Analysis* **4**, 21–30.

[69] Okajima, K.: 1998a, The Gabor function extracts the maximum information from input local signals, *Neural Networks* **11**, 435–439.

[70] Okajima, K.: 1998b, Two-dimensional Gabor-type receptive field as derived by mutual information maximization, *Neural Networks* **11**, 441–447.

[71] Papoulis, A.: 1984, *Probability, Random Variables and Stochastic Processes*, MacGraw-Hill.

[72] Peli, E.: 2002, Feature detection algorithm based on a visual system model, *Proceeding of IEEE 90*, Vol. 1, USA, pp. 78–93.

[73] Pellegrino, F. A., Vanzella, W. and Torre, V.: 2004, Edge detection revisited, *IEEE Trans Syst., Man, Cybern.Part B* **34**(3), 1500–1518.

[74] Petkov, N. and Kruizinga, P.: 1997, Computational models of visual neurons specialised in the detection of periodic and aperiodic oriented visual stimuli: bar and grating cells, *Biological Cybernetics* **76**(2), 83–96.

[75] Poggio, T., Voorhees, H. and Yuille, A. L.: 1985, A regularized solution to edge detection, *Technical Report Memo 833*, MIT Artif. Intell. Lab., Massachusetts Inst. Technol.

[76] Reisfeld, D.: 1996, The constrained phase congruency feature detector: Simultaneous localization, classification and scale determination, *Pattern Recognition Letters* **17**(11), 1161–1169.

[77] Robbins, B. and Owens, R.: 1997, 2D feature detection via local energy, *Image & Vision Computing* **15**(5), 353–368.

[78] Ronse, C.: 1993, On idempotence and related requirements in edge detection, *IEEE Trans. Pattern Anal. Machine Intell.* **15**(5), 484–491.

[79] Ronse, C.: 1995, The phase congruence model for edge detection in two-dimensional pictures: A mathematical study, *Technical Report Rap95-11*, Université de Strasbourg, LSIIT-URA 1871.

[80] Sarkar, S. and Boyer, K. L.: 1991, On optimal infinite impulse response edge detection filters, *IEEE Trans. Pattern Anal. Machine Intell.* **13**(11), 1154–1171.

[81] Shen, J. and Castan, S.: 1986, An optimal linear operator for edge detection, *IEEE Computer Vision and Pattern Recognition*, Miami, FL, pp. 109–114.

[82] Shen, J. and Castan, S.: 1992, An optimal linear operator for step edge detection, *Graphical Models and Image Processing* **54**(1), 112–133.

[83] Simoncelli, E., Freeman, W., Adelson, E. and Heeger, D.: 1992, Shiftable multiscale transforms, *IEEE Trans Information Theory* **38**(2), 587–607.

[84] Simoncelli, E. P.: 1994, Design of multi-dimensional derivative filters, *IEEE International Conference on Image Processing*, Austin, Texas, pp. 790–794.

[85] Simoncelli, E. P. and Farid, H.: 1996, Steerable wedge filters for local orientation analysis, *IEEE Trans Image Processing* **5**(9), 1377–1382.

[86] Tagare, H. D. and de Figueiredo, R. J. P.: 1994, Reply to "On the localization performance measure and optimal edge detection", *IEEE Trans. Pattern Anal. Machine Intell.* **16**(1), 108–110.

[87] Todman, A. G. and Claridge, E.: 2000, Low-level grouping mechanisms for contour completion, *Information Sciences* (1-4), 19–35.

[88] van Vliet, L. J., Young, I. T. and Verbeek, P. W.: 1998, Recursive gaussian derivative filters, *International Conference on Pattern Recognition*, Vol. I, Brisbane, Australia, pp. 509–514.

[89] Venkatesh, S. and Owens, R.: 1990, On the classification of image features, *Pattern Recognition Letters* **11**, 339–349.

[90] Weldon, T. P. and Higgins, W. E.: 1998, An algorithm for designing multiple Gabor filters for segmenting multi-textured images, *IEEE International Conference on Image Processing*, Chicago, IL.

[91] Weldon, T. P., Higgins, W. E. and Dunn, D. F.: 1996, Gabor filter design for multiple texture segmentation, *Optical Engineering* **35**(10), 2852–2863.

[92] Wiskott, L.: 1999, Segmentation from motion: Combining Gabor- and Mallat-wavelets to overcome the aperture and correspondence problems, *Pattern Recognition* **32**(10), 1751–1766.

[93] Young, I. T., van Vliet, L. J. and van Ginkel, M.: 2002, Recursive Gabor filtering, *IEEE Trans Signal Processing* **50**(11), 2798–2805.

[94] Ziou, D.: 1991, Line detection using an optimal IIR filter, *Pattern Recognition* **24**(6), 465–478.

[95] Ziou, D.: 1999, Finding lines in grey-level images, *Technical Report 240*, Université de Sherbrooke.

[96] Ziou, D. and Tabbone, S.: 1998, Edge detection techniques - an overview, *International Journal of Pattern Recognition and Image Analysis* **8**, 537–559.

In: Frontiers in Robotics Research
Editor: Max A. Denket, pp. 113-164

ISBN: 1-60021-097-X

Chapter 4

INTERNET-BASED INTERACTION WITH MOBILE ROBOTS

Alaa M. Khamis*

Department of Systems Engineering and Automation,
Carlos III University of Madrid, Spain
Avda. Universidad, 30, 28911 Leganés, Madrid, Spain

Abstract

The human interaction and the robot's autonomy are key functions that can spread the use of the robot in human daily life. The remote interaction is a special type of human-robot interaction, where the human and the robot are separated by physical barriers but linked via telematics technologies (= telecommunication + informatics). Due to the nature of intelligence needed for the modern robots, remote interaction systems must combine the dream of the artificial intelligence field of creating an intelligent being, creature, or program with human-computer interaction's vision of building systems that are centered around the user's needs and desires. This introduces a number of technical challenges, which must be addressed such as local autonomy, control strategy, communication medium constraints, feedback methods and control interfaces. In this chapter, these considerations are discussed. The chapter also presents a multilayer architecture, which can be used to facilitate the remote interaction with mobile robots using different interaction elements for research and educational purposes.

1 Introduction

Robots have seen an enormous innovation in very recent years to the point of entering our every day life. During the early sixties, when the industrial robots were first introduced in the manufacturing chains the potential of their use was expected to be very high even outside manufacturing. However, these expectations were proved to be false and numerous robotic applications were confined to research laboratories for many years despite the widespread use in manufacturing. Last decade, in particular, has witnessed intense research activities in

---

* E-mail address: akhamis@ieee.org

mobile robotics that covered all aspects ranging from design, sensorial fusion, control, environment modeling and mapping, to navigation and locomotion novel designs. Nowadays, robots have started a new era and entered new fields of application, including reaching our homes in the form of assistants in particular tasks, such as vacuum cleaning or even for the purpose of entertainment. There exist already commercial pet robots such as the AIBO of SONY and the robot PAPERO from NEC among others. Other advanced development also include the humanoid version of robots, being an example of that the HONDA robot ASIMO, which is able to walk up and down the stairs in a surprisingly agile manner. These developments were achieved owing to the recent advances in other technologies such as sensors, actuators, materials, new communications and multimedia technologies.

These new robotic trends are in their nature designed to interact more with the human user, whether for the purpose of entertainment or transmission of information for the benefit of performing given tasks (or service). This new reality entails the opening of new fields of research to deal with issues concerned with the human-robot interaction.

This chapter describes a software pattern-based architecture, which can be used to facilitate the remote interaction with mobile robots using different interaction devices. In the proposed architecture, a supervisory control strategy has been used in order to alleviate high communication data rates and system sensitivity to network delays. The users interact with the remote system at a more abstract level using high-level commands. The local robot's autonomy has been increased by encapsulating all the robot's behaviors in different types of skills. User interfaces have been designed using visual proxy pattern to facilitate any future extension or code reuse. Three types of remote drivers (PCs, PDAs & cellular phones) have been implemented based on this architecture.

This chapter is divided into seven sections. The following section describes the main components of remote interaction systems. The third section discusses how Internet robotics will change our life. The challenging problems of remote interaction systems are discussed in section 4 followed in section 5 by the proposed architecture. Section 6 presents a case study of the use of the developed system to provide remote experimentation for educational and research purposes and finally the chapter is concluded in section 7.

2 Remote Inetraction

The remote interaction is a special type of human robot interaction, where the human and the robot are separated by physical barriers but linked via telematic technologies. In recent years, human-robot remote interaction applications have grown in such way that they currently allow users with different levels of familiarity with robotics to perform different tasks efficiently. These tasks may be visiting museums, giving medical care for elderly people, creating artwork, navigating undersea, performing an experiment, testing new algorithms, controlling a household appliance, etc. However, regardless of the application, most of the remote interaction systems consist of the following basic components as shown in Figure 1.

Operator interfaces: incorporating an interaction device that the operator uses to send control commands to the remote system. There are many types of interaction devices such as PCs, mobile devices such as Personal Digital Assistants (PDA) and mobile phones, speech-based communication, joysticks and haptic interfaces.

Mobile robot system: that performs the operators' commanded actions at the remote site.

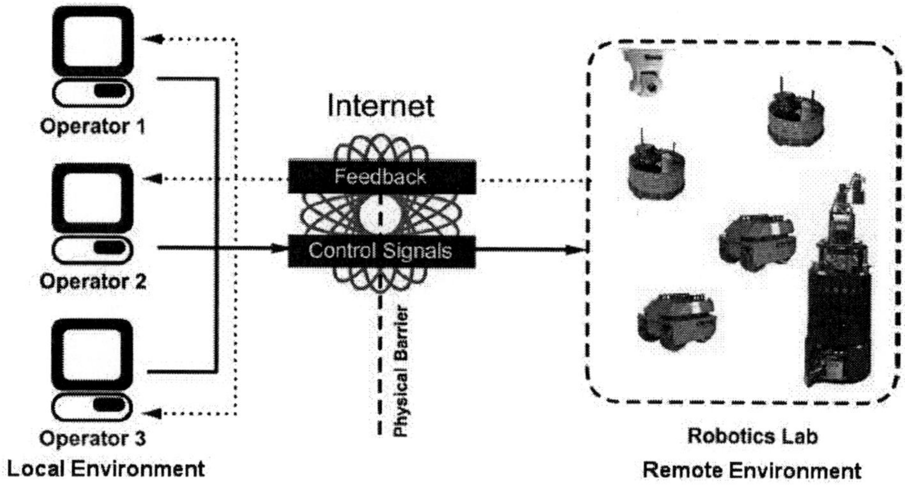

Figure 1. Remote Interaction

A communication scheme between sites: it is recommended that the communication scheme has a robust signal communication link with an acceptable time delay; dedicated data links with sufficient throughput; and an effective data loss-recovery approach. Although Internet is a cheap, readily accessible communication medium, its performance has nondeterministic characteristics.

Feedback interfaces: video transmission is commonly used to provide visual feedback for the operator. Video transmission demands high bandwidth availability. When this is not possible, computer-generated imagery supplies the operator with a virtual interface that combines low bandwidth sensory data to form a realistic image. These virtual interfaces are also useful to overlaying computer predictions or visual clues onto video images. Auditory feedback can also be used to enhance visual experience and human robot interaction. Although audio transmission requires low bandwidth, it is very sensitive to time delay and jitter. Other type of feedback can be provided using kinesthetic aids. Using haptic systems the operator can sense the response of its commands directly in the control interface.

3 How will Internet Robotics Change Our Life?

For the efficiency of any human-robot remote interaction applications, an efficient framework must lie underneath. During the last decade, several architectures have been proposed with different levels of scalability ranging from single operator single robot to multioperator multirobot interaction support. Most of these systems incorporate the Internet as a communication medium. These architectures are customly designed for various applications. This section briefly describes some proposed frameworks for a diversity of applications.

3.1 Space Exploration

The Web Interface for Telesciene (WITS) provides Internet-based ground operations for planetary lander and rover missions. It provides an integrated environment for operations at the central operations locations, collaboration by geographically distributed scientists, and public outreach [1]. As shown in Figure 2, the WITS system architecture includes a database, a server and multiple clients. The database is a structured file system that holds downlink data and uplink sequence information. The server provides communication between clients and the database. Clients are distributed over the Internet and provide the interface needed to view downlink data and generate command sequences. The WITS system is implemented using Java and some of its extensions, like Java3D (for drawing 3D synthetic models) and the Java Cryptography Extension (to assure secure communications) [2].

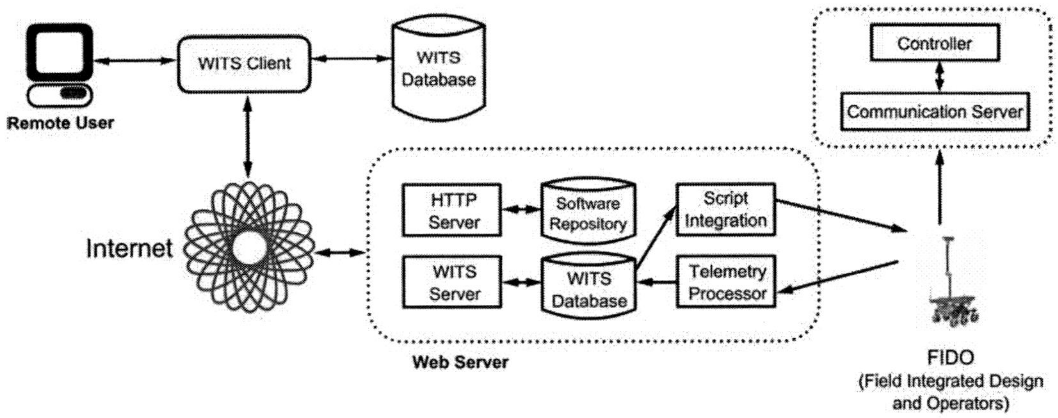

Figure 2. WITS System Architecture

This system has been used in the Mars Polar Lander mission ground operations for downlink data visualization and command sequence generation for the robotic arm and robotic arm camera and as a secondary tool for the stereo surface imager.

Furthermore, the Autonomous Robotic Systems (ARS) laboratory (University of Applied Science FH-Weingarten, Germany) has developed a system called RedRover for the European Space Agency (ESA) [3]. This robot can be controlled via the Internet to explore a simulated Martian environment at the university.

A distributed robotic system has been developed including a communication infrastructure between five robots and one human operator [4]. The tasks necessary to build this system include developing a wireless communication architecture, a communication protocol, creating new hardware for relative position signaling using infrared LEDs, integrating a wireless camera system with the robots, designing and building a graphical user interface for controlling the robots, and designing and programming a set of behaviors to facilitate exploration of an environment.

A different type of space application that was prototyped is the capturing of protein crystals on board a space station [5]. The purpose of this project is to investigate the possibility of allowing scientific operations to be conducted from earth via remote control. Scientists can therefore conduct their experiments and analysis on protein crystals without requiring astronauts to go there. For the prototype, controllable devices are located inside a physical mock up of the European Space Agency (ESA) microgravity glove box. Three video cameras are used to give visual feedback, as seen in Figure 3.

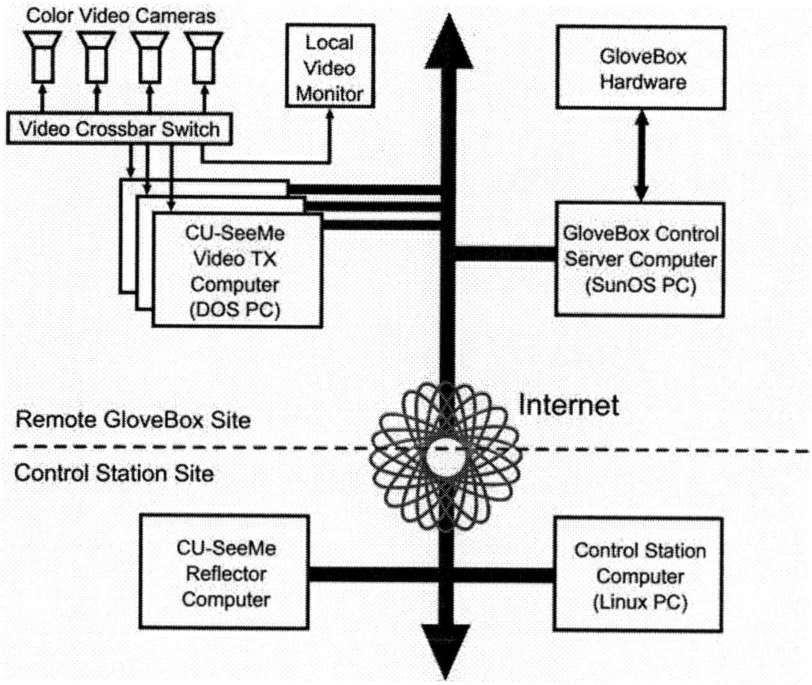

Figure 3. Communications topology used for capturing of protein crystals

There are two methods in which the user can control the device: using low-level commands and using macros which consist of sequences of low-level commands. For experimentation, the robot system was located in Washington, and the user control located in Texas. Experimentation was made for two different cases: communication through a dedicated link, and through the internet. Encouraging results were achieved, opening doors for the possibility of having a real remote operation in space over an Internet link.

3.2 Telemedicine

Internet-based telemedicine systems have been attempted by many researchers, and the most approaches are experimented with the use of videoconferencing for remote consultation and case study. However, current telemedicine applications include remote presence, telecare and telesurgery.

3.2.1 Remote Presence and Telecare

The "Companion" is a mobile machine developed by InTouch Health Inc. to give medical care for elderly people [6]. This system helps patients at a senior home to get personal care from a medical expert thousand of miles away. This system is not designed to replace a caregiver but rather to give remote presence expertise.

As shown in Figure 4, the care station is equipped with a joystick and flat screen monitor and has a camera attached as well. This gives the caregiver the ability to drive it down the hall and interact with patients. The use of this technology will enable greater efficiency and extend the reach of health care.

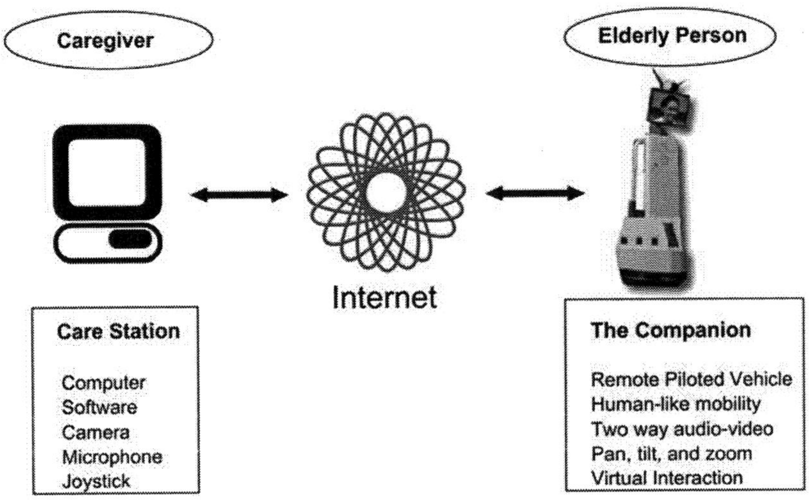

Figure 4. Telecare System

The concept of network distributed telecare robotic systems using Common Object Request Broker Architecture (CORBA) has been proposed in [7]. This healthcare platform allows the support group consisting of doctors, nurses, helper, and family members to develop their application services to aid the aged or disabled. It also has remote accessibility for hospital, healthcare organizations and relatives from any location where a Web browser is available. This system can realize delivery support and daily support such as providing tableware and food to the aged or disabled. The system also provides monitoring functions for all the supporters.

3.2.2 Telesurgery

In September 2001, the first major trans-Atlantic telesurgical operation (Operation Lindbergh) has been carried out between New York and the European Institute of Telesurgery in Strasbourg, France (round distance of more than 14,000 Km). To operate on the patient in France, a surgeon working from New York controlled the arm of a surgical robot. Computer engineers were able to get around that problem. The surgical robot was split into two parts, the control part in Manhattan and the operating part in France. Each part was hooked up to a computer. The two computers then talked to each other over a secure, fibre-optic digital telecommunications network--NOT the standard Internet, where data transmission times are notoriously unstable. The result was a time delay between the surgeon's movements and the return video image displayed on screen less than 200 milliseconds; noticeable, but manageable. The estimated safe lag time was 330 ms.

Siu et al. have attempted to develop a model, MAX, that can be used within an Internet-based mobile robot to autonomously open doors through the handicap button [8]. The architecture is shown in Figure 5. Two Java applets are used in the web interface, one responsible for direct control by the user, and the other to display video feed from the robot's CCD camera. A separate video server and control server are utilized for that purpose.

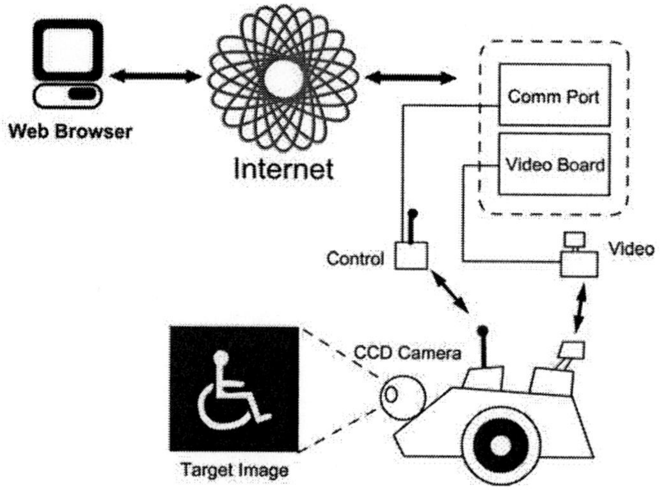

Figure 5. The MAX infrastructure

As another application of telecare, a distributed telerobotics system that provides simple services to the aged and disabled has been developed [9]. Components of the system include an omnidirectional robot, a robot arm, and a vision system. Each of the cameras in the vision system is used for a certain task: one camera recognizes objects through computer vision algorithms, another camera sends live images to the server as feedback, and three cameras localize the mobile robot. As shown in Figure 6, the robots and cameras are connected to a Windows NT4.0 server which also receives the client's commands through the Internet and sends useful feedback. It is noted however, that the three cameras used for robot localization are connected to an iGPS server before connecting to the main server. This is so in order to calculate the robot's position and orientation in real-time, which is quite processor intensive. The authors propose CORBA as a communication means to integrate between the different components of the system. The use of CORBA in this work alleviates variance in the components' interface language, operating system they are running on, and the network they reside at.

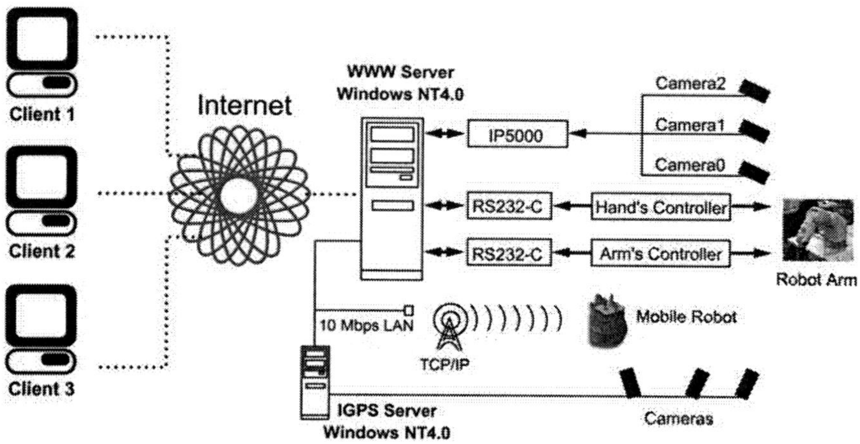

Figure 6. Robotic system to assist the aged and disabled

3.3 Interactive Artistry

The PumaPaint Project is an online robot that allows World Wide Web users to create original artwork [10]. The system hardware consists of a PUMA 760 robot equipped with a parallel-fingered pneumatic gripper. In developing this system, the ideas of an approach to time-delay teleoperation known as teleprogramming have been adopted. Teleprogramming has been shown to reduce and potentially eliminate the delay introduced by the "move-and-wait" approach. Furthermore, numerous systems have been developed to view artwork in a gallery or museum from anywhere in the world through an Internet browser. The Fisher Art Gallery at USC is such a gallery that allows real time viewing of 3D art objects through a pair of CCD cameras carried by a six degree of freedom robot arm [11]. Through a user friendly interface written in Java, the user can control the location and direction of the cameras within their workspace, as well as control the rotation of the rotary table on which the art statue resides.

3.4 Edutainment

Robots are used in education to support teaching robotics and other related concepts of mechanics, electronics, programming, etc. They are used in entertainment to implement games or support the production of movies providing robotic characters, or mechanisms to move cameras. Edutainment expresses the marriage of education and entertainment in a work or presentation such as a Web site or a system. Edutainment Robotics can be considered as the field where the most advanced ideas in robotics will be tested and put to operation. The Mercury Project allows users to command the remote arm to perform archaeological-like tasks [12]. The KhepOnTheWeb system allows the user to control a small Khepera mobile robot that carries a small color CCD camera and moves in a wooden maze [13]. A distributed architecture structured in four levels is also addressed, to enhance the performance of LEGO robots [14].

3.5 Remote Control of Household Appliances

Despite the phenomenal growth in wireless technologies and services in recent years there has been a relative lack of research on possible uses of these systems in the remote control of mobile robots. This mainly a consequence of the fact that the use of these systems has to deal with specific problems related to the limitation of handheld devices (slow processors, limited memory/storage, and small displays) and wireless network (restricted bandwidth, high latency, low connection stability, and low predictable availability).

PDA's are attractive interface devices because they are lightweight, extremely portable, and feature touch-sensitive displays. Most of the PDA-based interfaces have been written in PersonalJava and run under the Insignia Solutions JeodeTMJava Virtual Machine [15]. Because the interface requires high-resolution color, Windows CE-based PDAs are usually employed.

Fong et al have developed *PdaDriver*, a Personal Digital Assistant (PDA) system for remote driving [16]. PdaDriver uses multiple control modes to make vehicle teleoperation fast and efficient. A PDA-based controller has been designed at Cralos III University of Madrid to send direct control commands to the robot's actuator server and to acquire the odometry data (actual location with respect to the initial point, translational velocity and rotational robot velocity) [17]. Another interface has been implemented to remotely control the robot by using a Mobile Information Device Profile (MIDP)-enabled cellular phone. These cellular phones-controlled robots can open up new possibility for businesses looking to provide innovative in-home care and home security services.

In 2003, Fujitsu announced the beginning of commercial sales of a mobile phone-controlled, Internet-enabled robot called MARON-1 (Figure 7). This robot is an Internet-enabled robot that can be operated by mobile phone for remote household monitoring, remote control of household appliances, or to serve as a hands-free phone itself. With its onboard camera, it can transmit images of its surroundings to a remote operator using a mobile phone.

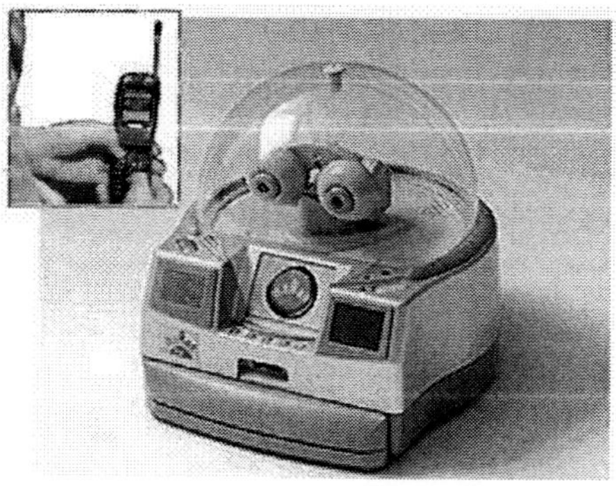

Figure 7. MARON-1

MARON-I can serve as part of an in-home healthcare communications system or a home security system for remote monitoring and notification in case of break-in. Such systems can be extended in future to be capable of delivering objects to and from people in the house, similar in concept to the office delivery robot mentioned in [18] and [19].

Fong et al have developed a multirobot remote PDA-based driving system with collaborative control [20]. They discuss how human-robot collaboration and dialogue provide an effective framework for making command generation and coordination efficient, so that human-robot interaction becomes transparent and tasks are easy to perform.

Internet Robots can even extend to applications such as laundry tasks similar to the work of Lung Ngai et al [21]. This robot can sort out laundry, and ask for human assistance whenever a problem is faced determining the color of a washcloth. The user communicates with the robot via an Internet browser. The robot then continues its task according the user's input. An Apache web server is used in this system to invocate CGI scripts that send information back and forth to the user. The web interface is written in HTML. A Java applet was used to continuously feed video frames to the user from several cameras that supervise the robot's actions.

3.6 Remote Inspection

For any autonomous robot vehicle, localization is an important attribute that must exist. That is the ability for a robot to locate its position within a given map. Two types of localization exist: local localization and global localization. With local localization, a robot can know its location after a transition, assuming that the initial (previous) location is known. This is achieved by assuming that the robot starts from a known initial position, and throughout the navigation the new locations are calculated using the displacement distances taken. Global localization is more complicated. Assuming that no a priori knowledge of a robot's initial location is known, the robot can investigate to find out where it is located on a given map.

Through global localization, robots can explore an environment for certain targets, as well as understand it and draw a map of it. This property is suitable for applications that require exploration of unknown environments, or of known environments but unknown robot location. Examples of such application include robot rescue operations (when people are trapped inside a building with the robot, the robot can be used as a means of communication or a means of delivering dole to the trapped people, in the case when the robot is small enough to pass through cracks), or even inspection of partially known environments (like ventilation systems or pipeline systems).

Alice (also know as the sugar cube robot) is one of few robots that were designed for such purposes [22]. It caught the attention of many people by its relatively tiny size, 2x2x2 cm, as well as its autonomous capabilities. Alice is capable of globally localizing itself within a given map. Alice can also build a map from its navigation path, but to only some extent. The architecture of Alice is fairly simple. A two-stage control strategy is used, where the first stage, local control, controls the motor, actuators, sensors, and external communication. It also supports autonomous operations, like obstacle collision avoidance, wall following, detecting another robot position, etc. The second stage, global control, concerns high level commands given by a user, like asking the robot to move in a certain direction, or along a wall. After the local control receives this command from the global control it would interpret it to the motors, and send feedback to the user. However, if an obstacle or dead-end is detected, the local control autonomously avoids the collision without external intervention.

3.7 E-Manufacturing Systems

Zhang and Wang have proposed a collaborative framework suited for controlling Shop Floor machinery through the web [23]. Java 3D models ware proposed instead of camera images which consume bandwidth. Figure 8 shows the architecture of this framework. The authors propose transmission of only sensor data and control commands between models and device controllers, a thin-client interface for navigation, and deployment of control logic in an application server.

Through kinematics modeling, parameters sent from a server controlling manufacturing robots display 3D behavioral models of the robots to the user. The goal of having this web-based architecture is to reduce traffic while providing realistic intuitive environments to be viewed by the users.

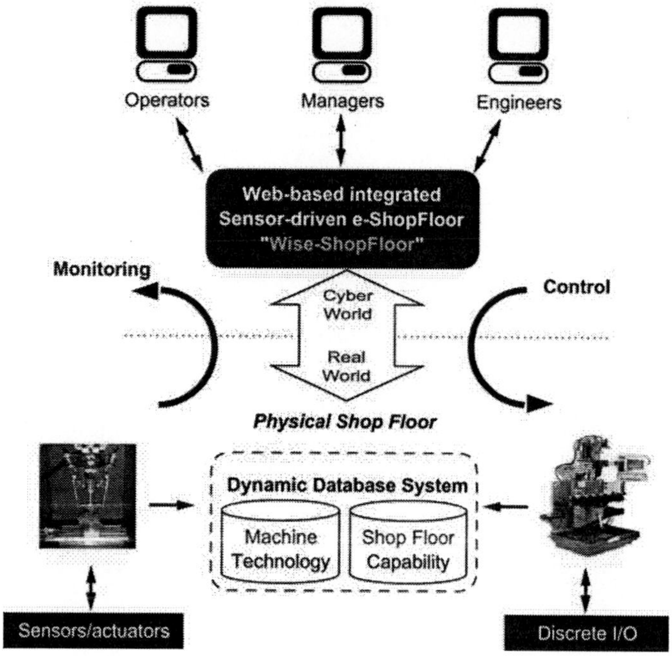

Figure 8. The Web-based digital Shop Floor framework

3.8 Foraging Applications

Foraging is a term used to describe the real world behaviour of searching for and analyzing an unknown environment. Similar to remote inspection, multirobot foraging helps the multirobotic system to localize its location globally.

A new framework for remote cooperation of multiple online robots is experimentally simulated in the application of multirobot foraging [24]. This agent-based architecture, uses a Java web interface similar to the skeleton in Figure 9, to let the user control the robots either through direct control or through high level behavioral commands. An interesting scheme used in this system is that whenever the central computer server fails, control is moved to one of the robots, where it starts operating as the host computer. If that robot fails, control is moved to another robot, and so on and so forth. This encourages autonomy and fault tolerance. This architecture is simulated and demonstrated on robot foraging behaviors. The forage command given to robots is divided into three tasks for each robot. First, the robots wander around randomly in a closed area searching for attractors. The next step is to approach the detected attractors, then finally, to deliver the attractor back to the home base.

As another example of Internet-based single operator multirobot systems with foraging objectives, the study of animal behavior (ethology) and the underlying neural structure responsible for behavior (neuroethology) is incorporated to model prey acquisition and predator avoidance neuroethology behaviors [25]. These behaviors are useful to prevent inspection robots from entering a precarious path or approaching a dangerous object. In this work, an architecture is constructed to embed mobile robots to distributed computational resources.

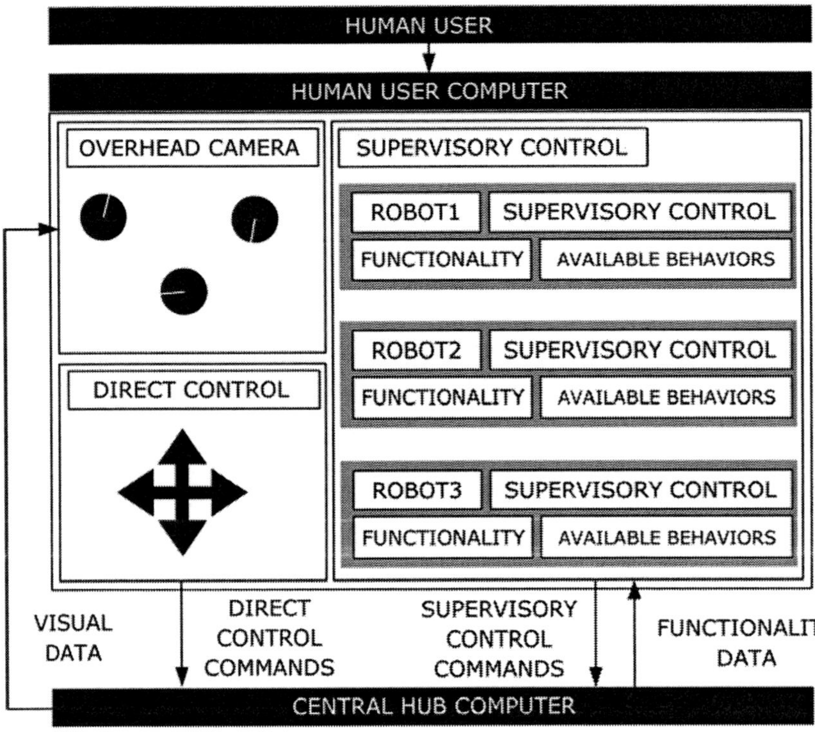

Figure 9. User interface for a multirobot foraging system

3.9 Hazardous Tasks

By 2010, approximately 300 nuclear power plants will require decommissioning [26]. This requires characterization of harmful and polluted environments. Therefore, a safe and efficient technique of doing so is highly demanded. Remotely interactive Internet robots would serve an excellent cause in decommissioning power plants. A robust and efficient architecture that can deal with the irregularities of the internet is required.

Cragg and Hu describe a mobile agent-based architecture suitable for nuclear decommissioning [27]. The architecture is suited for multirobot operations controlled by the user through the Internet. The advantage of the proposed architecture is the high fault tolerance of allocation of tasks to members of multiple robot teams. A fault-tolerant software architecture, ALLIANCE [28], is employed within the system. Fault tolerance is the multirobot team's ability to respond to individual robot failures or failure in communication that may occur at any time during the mission [27].

Figure 10 shows the architecture, divided into two layers: the user computer layer and the robot layer. Two agents reside on the user layer: the control agent, and the mapping agent. The control agent is responsible for task allocation from the user to the robots.

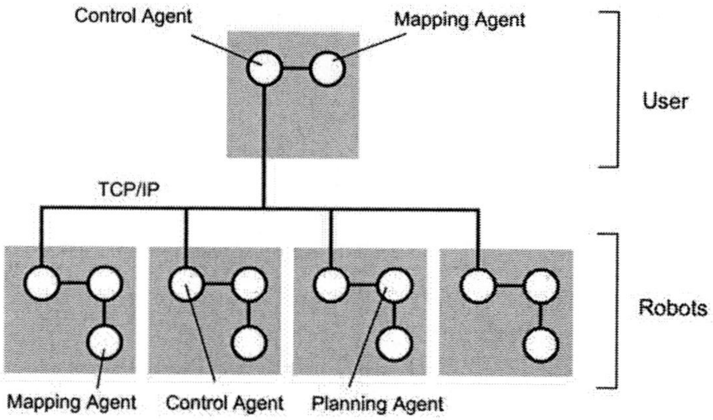

Figure 10. Architecture proposed by Cragg and Hu [27]

This task allocation can take of a number of forms: low level, medium level, and high level commands. Low level commands consist of direct control of the robots from within the user interface. Examples are to move the robot in a certain direction or rotate to a certain angle. Medium level commands consider taking a sequence of low level commands from the user and sequentially running those commands on the robot. High level commands, which have shown the best results, execute one of four different autonomous behaviors. Behaviors include: Mapping (maps and explores the environment), recharging (the robot returns to recharge from a recharge station), homing (the robot returns to a home location), and stationary (the robot remains stationary). Each robot contains a control agent to enable robot to robot communication.

The planning agent makes decisions (e.g. the direction that the robot should explore), based on its short term and long term map. Additionally, this agent determines the best time for a robot to depart back to its recharging station for a recharge, taking into consideration the power to be utilized in order to return to the station. Reinforcement learning is applied here. The mapping agent is responsible for construction of a long-term map used by all of the robots through gathering robot status and short-term map information from each of the robots. Bayesian or Dempster-Schafer Sensor Fusion Algorithms can then be applied. This agent allows the robot system to be used in more than one nuclear plant, disregarding the environmental map. Although the system was not tested through the Internet, a simulation of Internet performance was made by controlling the time delay of a local network.

3.10 Sharing Control

The first collaboratively controlled robot on the Internet system was developed by Goldberg et al [29]. This system allows a distributed group of users to simultaneously teleoperate an industrial robot arm via the Internet. A Java applet at each client streams mouse motion vectors from up to thirty users; a server aggregates these inputs to produce a single control stream for the robot. Users receive visual feedback from a digital camera mounted above the robot arm. Furthermore, Goldberg et al in another research project continue to consider systems where a group of humans share control of a single entity [30]. In this special kind of application, remote users can send commands to a skilled human, known as a tele-actor, who navigates and performs actions in an environment. The architecture of the system is shown in Figure 11. The authors propose using a real human instead of a mobile robot due to the problems that may inflict in terms of mobility, skill, and power consumption.

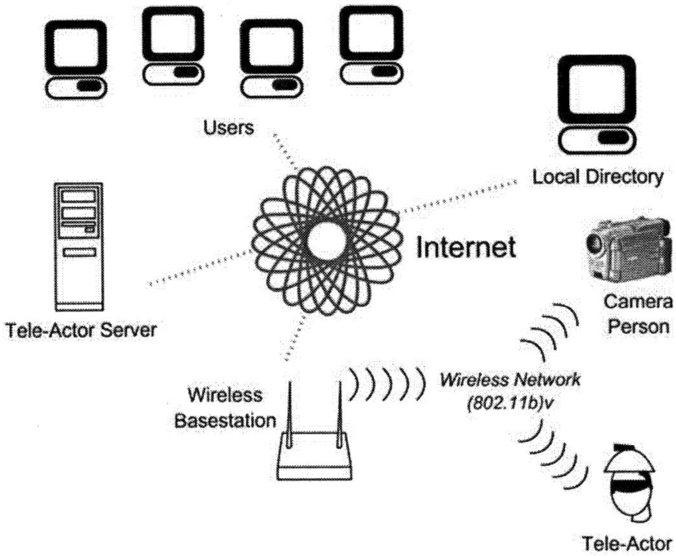

Figure 11. The tele-actor system architecture

For different circumstances, the participants are provided with a different voting image, which is a single image with a textual question. The participants click on their screens to vote for the most appropriate answer. The positions of the clicks are sent back to a server and to the browser interface, displaying all participant decisions to the users (providing voters the option to change their votes). Spatial Dynamic Voting (SDV) analysis algorithms analyze the votes to outcome a single control command, which is sent to the tele-actor.

It was noted that the participating students would often become passive in participating, and would only watch the video of the tele-actor. Furthermore, the experiment assumes that the choices of output are known, and hence input is compared with desired output.

It is not always the case that output possibilities are specified in advance by an external supervisor. To deal with these two problems, Song and Goldberg describe an unsupervised learning approach, where the desired outputs are unspecified [31]. Using this approach, participant scores are efficiently measured. The experimentation was carried out using the classic twister game. Teams of online players direct the stances of two human players in real time. The aggregated output that determines the next move for the player is based on the majority cluster of consensus region of all votes. Participant scoring is based on active participation and collaboration.

3.11 Remote Surveillance

Robotic video cameras that can pan, zoom, and tilt are now very common for use in surveillance systems. For Internet-based surveillance systems, a single user can control a set of cameras to give a desired view. The trend today is to allow the potential of multiple users accessing the same surveillance system; hence a multioperator single robot system.

An example is ShareCam, a robotic web-based camera that accepts many requests from online users [31]. As shown in Figure 12, the ShareCam system consists of the ShareCam camera and two servers that communicate with online users. Users interact with the system via a Java applet. In the applet, a panoramic image is shown, representing the focal range that can be viewed by the camera. Online users depict frames (on the panoramic image) they would like the camera to view. The system computes the optimal camera view, servos the camera, and provides the video feed to all users.

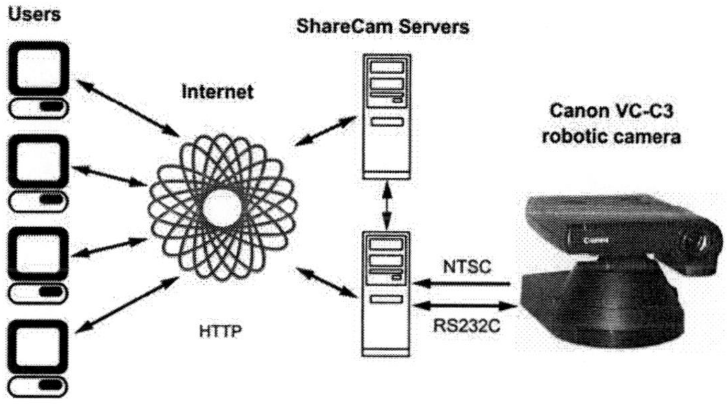

Figure 12. ShareCam architecture

3.12 Remote and Distributed Laboratories

Remote laboratories can be considered innovative environments, which can be used to provide remote interaction with mobile robots for educational and research purposes. Remote laboratories can be defined as network-based laboratories where the user and the real laboratory equipment are geographically separated and where telecommunication technologies are used to give users access to laboratory equipment [32]. Such laboratories have the advantage that they are not restricted to synchronized attendance by instructors and students: thus they have the potential to provide constant access whenever needed by students.

In recent years, many researchers have built remote laboratories for mobile robots using similar architectures and implementation tools [33]. Many such remote facilities can be put together to form a framework or a distributed laboratory that can be used to provide a coordinated set of experiments for students with hardware facilities physically spread over different locations, but accessible via the Internet as shown in Figure 13.

The project IECAT (Innovative Educational Concepts for Autonomous and Teleoperated Systems) in which we have participated is an example of such frameworks in the field of mechatronics [34]. Such frameworks aim to be an electronic workspace for distance collaboration and experimentation in research or in another creative activity, to generate and deliver results using distributed information and communication technologies. They assist in the exchange of existing hardware resources and educational materials between the partners.

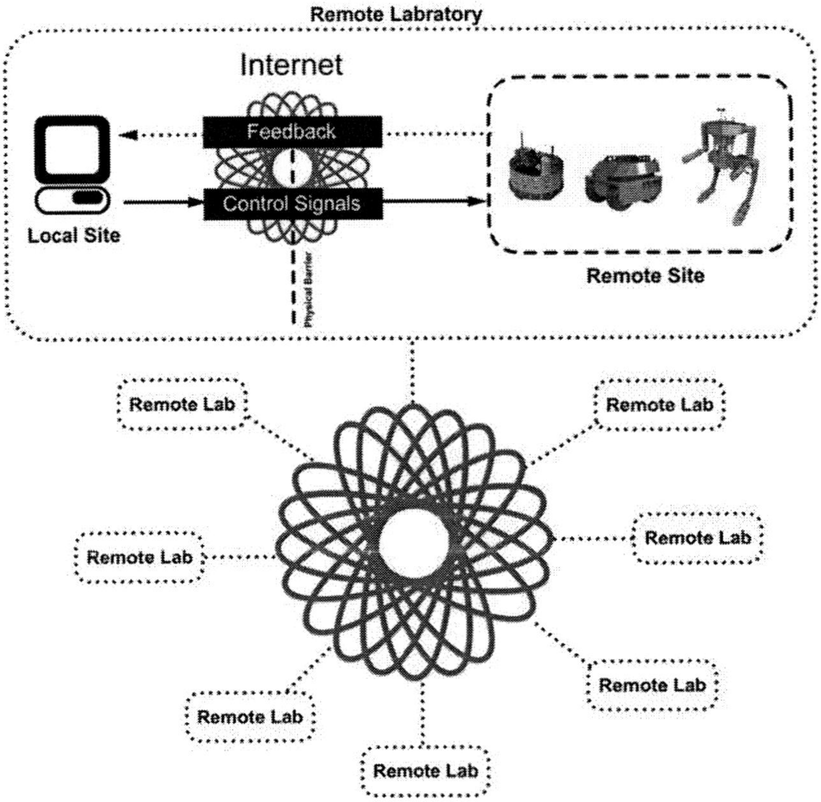

Figure 13. Remote and distributed laboratories

In the designing of these distance laboratories for robotic systems, a number of challenges must be addressed, particularly the telematics infrastructure which gives access to experiments, as well as the user interface which provides the necessary interactivity with the remote hardware supporting the learning process of students through appropriate feedback.

Gallagher and Perretta show the clear advantages of using the Web to provide robot access as a simulation tool open for students [35]. As the popularity of courses in the field of design and programming of mobile autonomous systems is increasing at a number of institutions, it is very beneficial for students to test and practically see their programming design efforts. It may also motivate students to continue in such a field. They have used the Java Khepera Simulator in their WWW course [36].

A similar architecture is proposed that surpasses the latter in the flexibility to manage different robots or include new robots and objects in the system without needing to modify the user interface or the architecture (shown in Figure 14) [37]. The intended purpose is to provide a training environment for students to carry out practical experiments to understand kinematics and trajectory design for industrial robot arms by teleoperating the robot through a browser. The system displays a realistic simulation of an industrial robot arm through the browser using Java 3D technology. However, it is not explained in detail how the system collaborates between different users who are accessing the robot at the same instance.

The design and the implementation of remote experiments on motor control of mobile robots are addressed in [38] with particular emphasis on user interface enhancement through the use of virtual reality models, as well as on network performance tests.

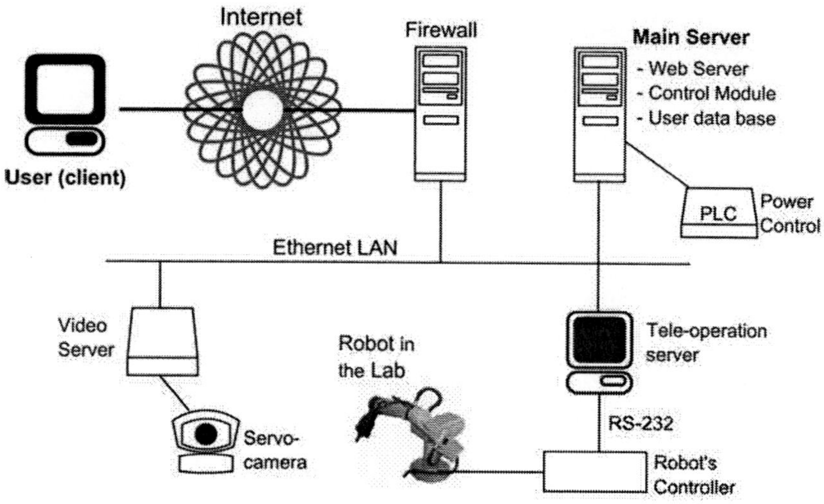

Figure 14. Architecture used for simulating and tele-operating an industrial robot

Figure 15 shows the TOM (TeleOperated Machine) architecture. The client-level interface includes position and speed control by adjusting the left and right wheels voltages, as well as the movement duration. Also the client can study the effect of the PI controller parameters on the robot. Round robin approach has been implemented to manage the user access. When only a single user is connected, no restrictions on usage are given. However, if multiple users are simultaneously connected, only one is given access for a specified amount of time. Once the time has expired, the next user in line will be given opportunity to control the robot and the first user will be moved to the end of the waiting list. A chat room is provided as a communication mechanism between the different users. Also a web cam with a streaming server provides a real time view of the robot movements. Through the use of a virtual reality user interface the operator can achieve a global view of the experiment state, in a fast and efficient way.

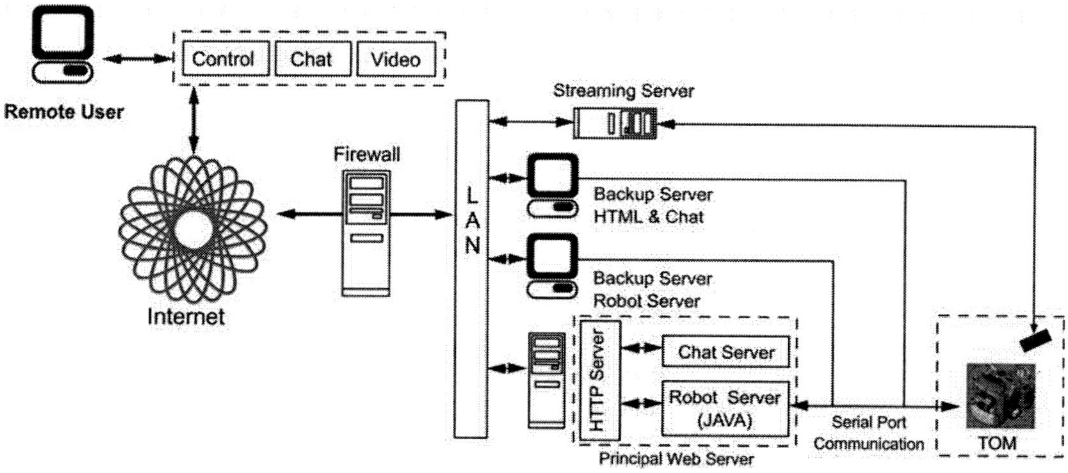

Figure 15. The TOM Architecture

3.13 Multioperator Multirobot Experimentation

Internet-based multioperator multirobot systems are the most complex of all robotic remote interaction systems. Currently, it is not feasible to use this type of system in applications with the current nondeterministic attributes of the Internet. The Internet creates a bottleneck for the capabilities of this system. However, some unique experimentations have been done to test these systems on certain applications, using the Internet as the communication medium.

One of these experimental applications is one where workers control robots to collaboratively perform an operation, while the robots autonomously cooperate in order to meet some criteria. Based on this kind of application, a system and a theory to control the force exerted by two robot manipulators on a metal rod is proposed, using an introduced concept of coordination of multiple robots [39]. For the experimentation phase, the robots, which are located in Michigan State, were controlled by operators located in Hong Kong and Japan. The manipulators coordinate autonomously in order to avoid exerting too much force on the rod from one operator.

Other applications served for experimentation of Internet-based robotic multioperator multirobot systems are multiplayer games. As an example, a multioperator multirobot system is proposed, where Internet participants teleoperate a pair of robots to perform a complex operation: play with a Rubik's cube [40]. Three control modes are provided for each user: principal operator, which gives full access to an individual robot, participant, which gives collaborative access with other internet users, and audience, which allows the user to observe the demo and game playing programs. Several experiments were made to demonstrate the system's capabilities. The first experiment involved mapping basic robot motion instructions into a set of teleoperation commands to be used by Internet clients.

The second experiment involved delivering an object from the first robot to the last robot through the three middle robots (five in total). The robots autonomously cooperate with each other to grasp or release the object efficiently. The third experiment involved testing the multioperator multirobot mode of the system. Two fine manipulators hold the Rubik's cube from opposite sides, and accept functions such as turning, changing the grasp, and alignment. Force/torque measurements from sensors were used as thresholds to determine whether the task has been done successfully. It is not discussed how the system manages and aggregates multiple user inputs.

A distributed architecture based on the usage of intelligent user interfaces and multi-agent systems has been described in [41] to facilitate the cooperative remote interaction with a multirobot system. In this architecture, a community of cooperative agents is assigned to provide different modes of remote interaction with a multirobot system.

4 Challenging Problems of Remote Interaction

To develop a remote interaction system, a number of technical challenges must be addressed such as local autonomy, control strategy, communication medium constraints, feedback methods and control interfaces. The following subsections discuss these considerations.

4.1 Communication Medium Constraints

Although the Internet provides a cheap and readily available communication medium for remote interaction systems, there are still many challenges that need to be solved before successful real-world applications can be achieved. These problems include restricted bandwidth, random time delay and data loss, which influence system performance. Internet performance as a communication medium has nondeterministic characteristic, which depends mainly on the network load. The performance of a computer connection is based on the speed and reliability with which data are transmitted over that connection. The speed and reliability of the entire Internet cannot be measured at present. Very little quantitative performance data is available, but a number of projects have been created recently to analyze performance. Currently, the Internet Traffic Report [42] and Internet Weather Report [43] provide an approximate measure of Internet performance by measuring the performance of connections among a small group of monitored sites distributed throughout the world. Delay and packet losses are monitored as measures of connection speed and reliability respectively. Other projects are trying to develop metrics to analyze overall Internet delay and packet loss such as the Internet Performance Measurement and Analysis Project [44] and the Internet Protocol Performance Metrics [45].

Many researchers have studied the use of Internet as a communication medium from different points of view. Brady and Tarn discuss Internet-based remote teleoperation including a theoretical foundation for modeling communications delays [46]. Han et al have proposed a control architecture, which guarantees that a personal robot can avoid obstacles and reduce path error and time difference between a virtual robot at the remote site and a real robot at the local site. The proposed architecture is insensitive to the inherent time delay where the personal robot is controlled using a simulator provided at a remote site [47].

Others try to use the Quality-of-Service (QoS) model to improve the efficiency of Internet-based systems [48,49]. On the Internet and in other networks, QoS is the idea that transmission rates, error rates, and other characteristics can be measured, improved, and, to some extent, guaranteed in advance [50]. QoS is usually described by four parameters, time delay, bandwidth, jitter and packet loss. The implementation of QoS model in today's networks is still in process, thus giving the opportunity to propose new dynamic tuning parameters [51]. Internet-based systems can take advantage of the implementation of the QoS model not only through the guaranteed end-to-end behavior, but also by the control of network parameters through a performance criterion in order to adapt the performance to the task requirements.

Other researchers study the Internet effects from user perspectives. A quantitative evaluation of operability that depends on communication time delay in teleoperation and time perceptions of human operator presented in [52]. The experimental results of this study showed some tendency that from 0.6 sec to 1 sec in delay there was some change of operator's operation time. Other qualitative studies show that people seem to be able to compensate for (learn) small added delays, but cannot learn large ones (>100 ms.) and the operator performance decreases significantly as the time delay increases beyond 0.2 s [53].

A survey was conducted to evaluate the network performance between the University of Applied Sciences (FH-Weingarten) in Germany, and the University Carlos III in Madrid (UC3M), which are partners in the IECAT project. The results of the study have been used as guidelines for selecting the suitable control strategy for the remote laboratory. Many tools can be used to measure the network performance such as ttcp [54], pingER [55] or hp netperf [56]. In this study the Netperf tool was used, which consists of two different executables: the netperf client in a 200MHz Linux Machine at FH-Weingarten, and the netperf server installed in a 300MHz PIII Linux Machine at UC3M. The test was re-run several times using different test durations and it was being run two times per day (at 11 AM and at 6 PM) and the average was calculated to insure high accuracy.

4.1.1 Time Delay

TCP Connect/Request/Response Test has been used to measure the round trip time between the two sites. This test mimics the http protocol used by most web browsers. Instead of simply measuring the performance of request/response in the same connection, it establishes a new connection for each request/response pair. The measured round trip time has been used to calculate the total time delay between the client and server based on the TCP analysis model described in [57]. Figure 16 shows a comparison between the local and remote host time delay, which is the time elapsed from when the client (at Fh-Weingarten) initiates a TCP connection until the client receives a requested 10kB web page stored in UC3M server.

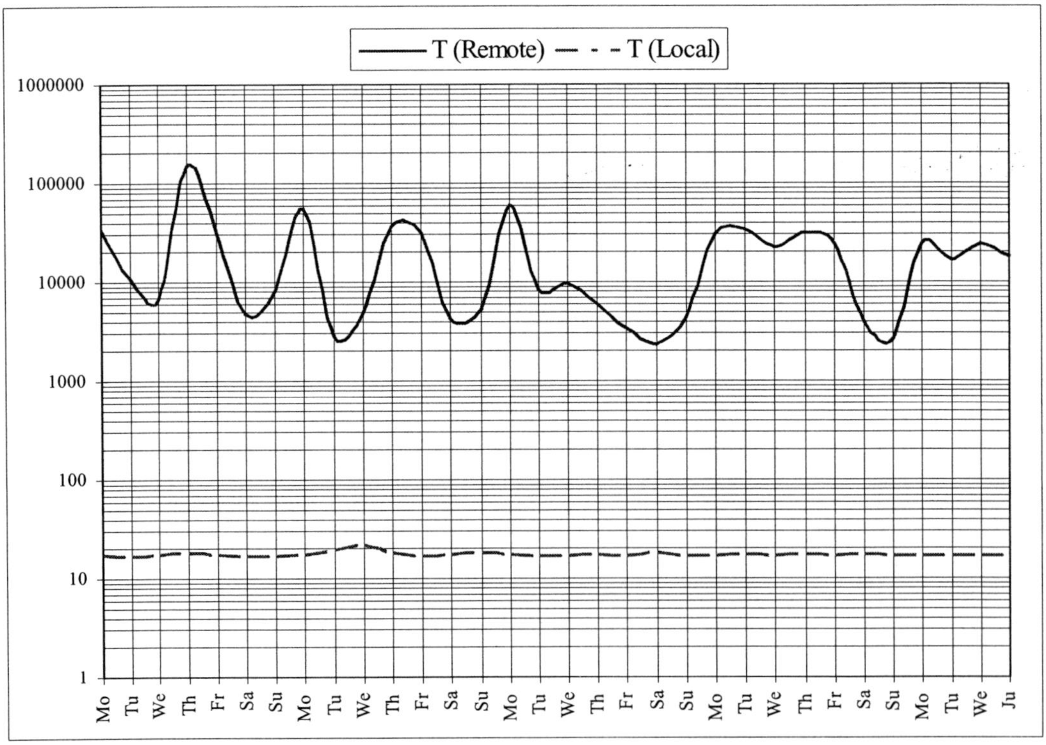

Figure 16. Time Delay Variation in ms.

The results show that the Internet imposes serious time delays, which are highly unpredictable and inevitable, unlike traditional teleoperation systems where a dedicated transmission link is used to guarantee the time delay and the bandwidth.

Another study has been carried out to compare the response time from different hosts. Figure 17 shows the results of the comparison between the average response time calculated during one week from FH-Weingarten, Germany as local host and four remote hosts, which are UC3M, Spain, University of Aalborg, Denmark, Helsinki University of Technology, Finland and Utah State University, USA. The used model supposes that a 10 Kbytes html file with one referenced object be transferred from the remote host to the local host.

The results show that the physical distances between the hosts have not dominant effect on the time delay. The Internet time delay is caused by the queuing delay, the processing delay and the transmission delay in the switches and the propagation delay in the links. The queuing delay defines the time; a packet waits in the buffer of a switch for transmission onto the next link. Therefore its value varies with the network load. The propagation delay depends on the physical distance due to the speed of light. With regard to all other delays adding to the total communication delay, the queuing delay represents the major portion. The queuing time depends mainly on the network load, which causes the arbitrary feature of the delay. The Internet delay variation is the most important problem to handle when implementing remote interaction systems. Depending on the network load, the delay can vary from a few milliseconds to hundreds. This delay affects the reliability of the remote operation. Beyond a certain delay, manual control of a vehicle may become highly error prone or impractical [58]. A maximum time delay of 1 sec is usually taken as a reference of operability in remote control systems.

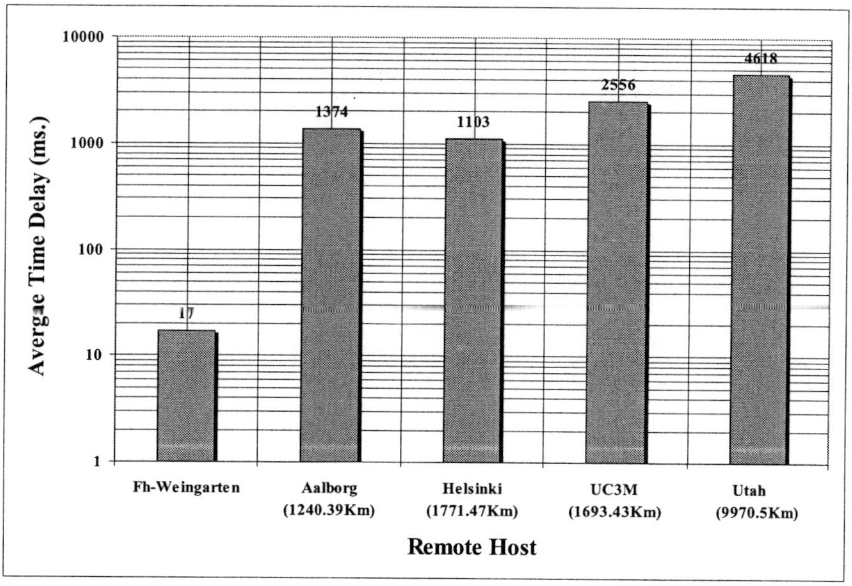

Figure 17. Average Time Delay measured from Different Sites.

4.1.2 Bandwidth

Bandwidth has a general meaning of how much information can be carried in a given time period (usually a second) over a wired or wireless communications link. In data transmission, *throughput* is the amount of data moved successfully from one place to another in a given time period [50]. To evaluate the bandwidth between UC3M and FH-Weingarten, the Netperf Stream test has been used. In TCP Stream Test netperf client sends messages of selected size to the netserver daemon which receives them. There is no protocol beside the TCP/IP protocol but all messages sent are also considered as correctly received hence the term throughput can be used for the bandwidth. A timer makes the test stop and the throughput is calculated in both sides by using the message size, number of messages and elapsed time. The test was re-run several times using different test durations and the average throughput was calculated. Figure 18 shows the results of this test.

In [59] a teleoperation system for the mobile robot Nomad has been presented, which gives us the opportunity to operate Nomad safely from the distant control centers. Images and data from Nomad were also immediately available on the Internet. Bapna et al have proved that the 1.4 Mbps bandwidth is sufficient enough to transport real imagery from the rover to a local control station and then to remote mission control sites. The obtained results (average=1.61 Mbps) can be considered satisfactory but it is necessary to take into account the nondeterministic feature of network performance.

In Internet-based systems, high-level commands are recommended to be used in remote interaction with the mobile robot because these commands require less bandwidth. The idea - of overcoming communication constraints by communicating at more abstract level and increasing the robot's autonomy- is fundamental to remote control via constrained communications.

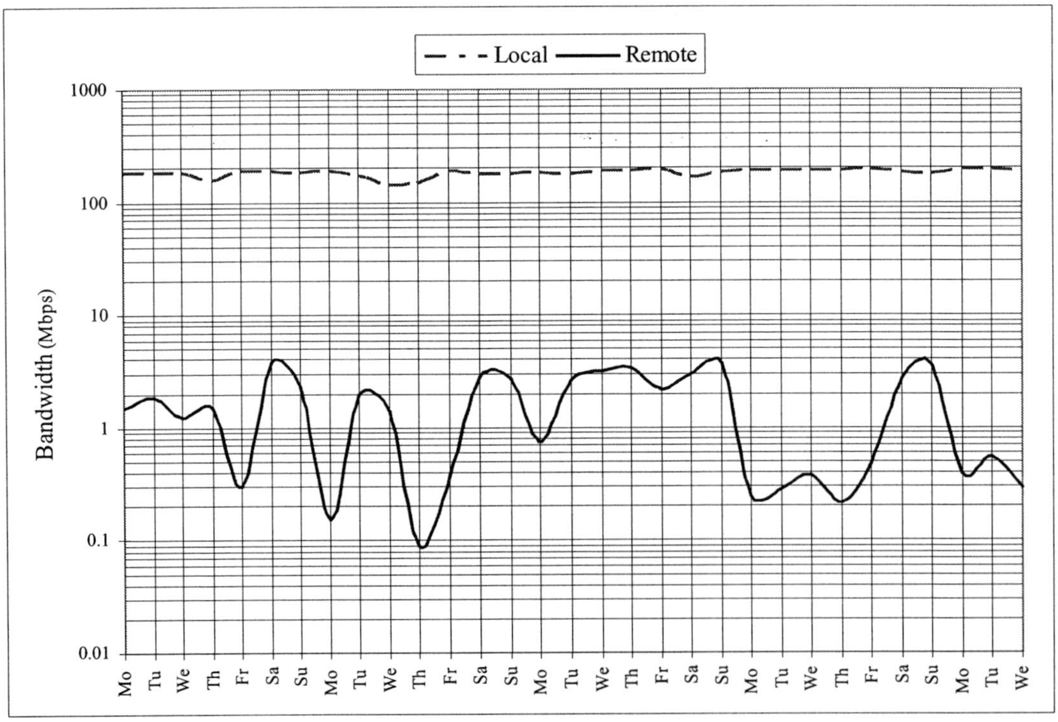

Figure 18. Bandwidth in Mbps

The limitation in bandwidth also limits the refreshing rate of the video images, which prohibits experiments in which the scene might vary at high speeds. Otherwise abrupt breaks and jumps in the visualization of the events will occur. Virtual reality images of a rendered simulation have much lower needs in throughput; therefore they pose a better alternative to real camera images in these kinds of situations [38]. If real time images must be provided from remote sites, the system should react dynamically to changing bandwidth and computational resources and only transmit those pixels that are actually needed by using intelligent techniques (intelligent fragmentation, intelligent frame rate, intelligent task rate and brute force compression) described in [60]. In multi-user systems, system resources have to be shared and some types of interaction should be detrimental or prohibitive, particularly those that require a high and dedicated communication bandwidth.

4.1.3 Jitter

A crucial component of end-to-end delay is the random queuing delays in the network devices. Because of these varying delays within the network, the time elapsed from generation of a packet at the source until it is received can fluctuate from packet to packet. This phenomenon is called instantaneous variability or jitter. Jitter is the variance in one-way latency. The previous round trip measurements can be used to measure the jitter. Let the i-th measurement of the round trip time (RTT) be R_i, then the "jitter" is taken as being the Inter Quartile Range (IQR) of the frequency distribution of R [61]. From descriptive statistics of the measured round trip time, jitter can be calculated as following:

Jitter = Inter Quartile Range (IQR) = third quartile (Q3)- first quartile (Q1)= 1911-340= 1571 ms.

This value represents high jitter, which can negatively affect the performance of data transmission especially in speech applications. Although speech transmission does not need a high bandwidth, it is very sensitive to delay changes. Gateways can be used to delete jitter in such applications. Many mechanisms have been proposed to remove the jitter in some special applications such as voice applications or Internet telephony [57].

4.1.4 Packet Loss

Package loss is caused by exceeding the network capacity causing a network device to drop a packet. Probably the greatest concern of any Internet-based systems is the nondeterministic system behavior that would result during packet loss or total dropout. One possibility in order to prevent package loss is implemented in TCP, where, as soon as package loss is detected, a resend is requested by the receiver. That results in higher latency regarding TCP compared to UDP, so there is a trade-off between the parameters package loss ratio and time delay [51].

4.2 Control Strategy

The control strategy of the remote interaction system plays an important role in the performance and stability of the system. When there is significant time delay in the communication link, instability occurs and manual closed-loop control is no longer suitable [62]. Two control strategies are discussed here, which can be used to develop Internet-based systems.

4.2.1 Manual Closed Loop Control

Most of the Internet-based systems use the manual closed loop control model because of its simplicity. These systems tolerate delays by using the human operator as an intelligent link to the control system. An error in the position of the remote device is visible in the returned imagery, and the operator sends commands to the remote device to compensate. This model has the components shown in Figure 19.

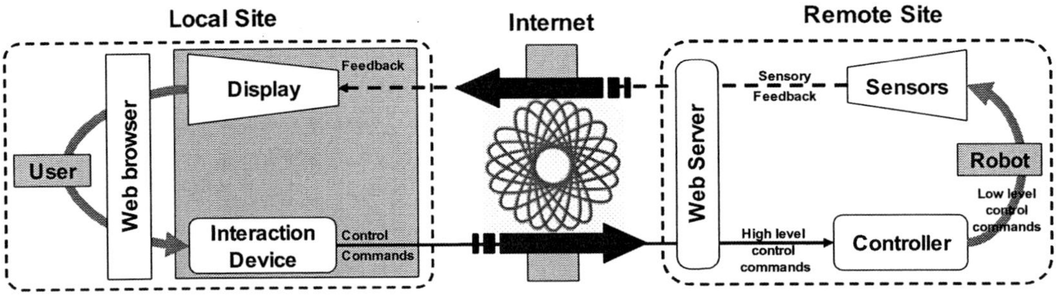

Figure 19. Manual Closed Loop Control Model

It is based on a simple protocol commonly used in distributed computation "The Request/Response Protocol". The client interacts with the system using any Web browser to make the request. Client requests are translated into HTTP requests, which are satisfied by the Web server. These requests are converted to high-level control requests that are received by the robot controller, which transmits them as low level control requests to execute the required task. Sensory feedback is required to give the user information about the remote robot's environment and the consequence of his/her commands. While these systems are not certainly optimal, they are simple and they work. Predictive displays can be used to improve this model. Because of the communications delay, immediate feedback cannot come from the remote site, and instead it must be generated at the operator station [60].

4.2.2 Supervisory Control

One technique that can improve operator performance and avoid the instability problem is a shared or supervisory control scheme, where control of the robot is shared between a remote control loop and the local human operator [62]. The goal is not to make a robot that can realize the full task autonomously but to enable the robot to perform some simple subtasks that the operator can sequence. By supervisory control, the user can communicate with the remote system at a more abstract level using high or task-level commands sent to a robot, which has an increased level of autonomy in order to alleviate high communications data rates. Limiting the remote interaction to high level commands helps to decrease the bandwidth needs and increasing robot’s autonomy can help to decrease teleoperated task sensitivity to the delay. Figure 20 shows the main components of the control model used to develop the remote laboratory.

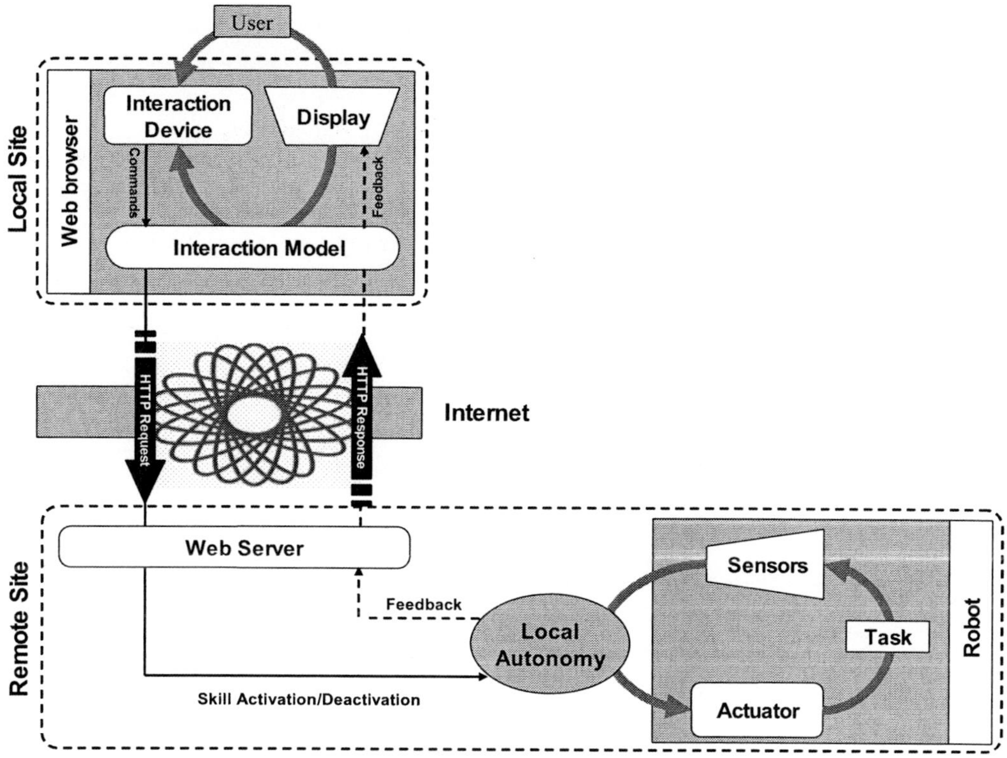

Figure 20. Supervisory Control

4.3 Robot Autonomy

To increase the local autonomy of the robot the concept of skill has been used. A skill represents the robot's ability to perform a particular task. They are all built-in robot action and perception capacities [63]. The AD (Automatic Deliberative) control architecture has been proposed to develop different robotic skills [64]. As shown in Figure 21, this architecture has two main levels.

The automatic level contains low-level control modules, which act directly upon the actuators, as well as the modules that collect data from the different sensors of the system. The automatic skills have been classified into motor, perceptive and sensorimotor skills. Motor Skills are those skills which generate actions for the robot's actuator such as direct control skill. Perceptive Skills are those skills which interpret the information obtained from sensors or other perceptive or sensor-motor skills. Sensorimotor Skills obtain as input the information provided by sensors or other perceptive skills, and based on this information choose the most adequate actions for the actuators.

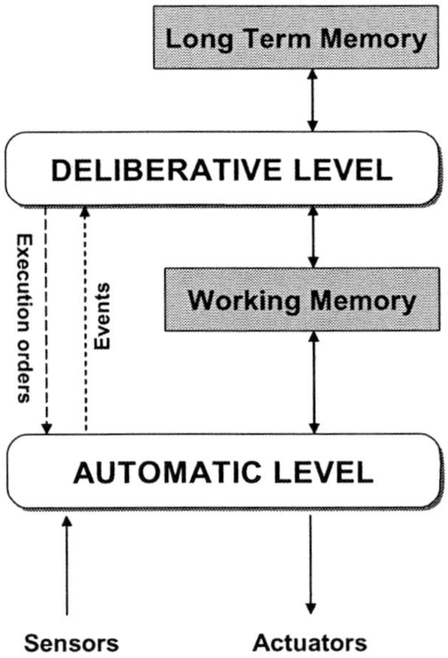

Figure 21. AD Architecture

The other level is deliberative, which contains modules that require reasoning capacity. Those modules do not produce immediate responses. They need a longer time to process the information they work with. Those modules form the deliberative skills such as path planner, the environment modeler and the task supervisor.

4.3.1 Skill Structure

Skills have been implemented in form of client-server modules [65,66]. As shown in Figure 22, each module contains an active object, an event manager object and data objects.

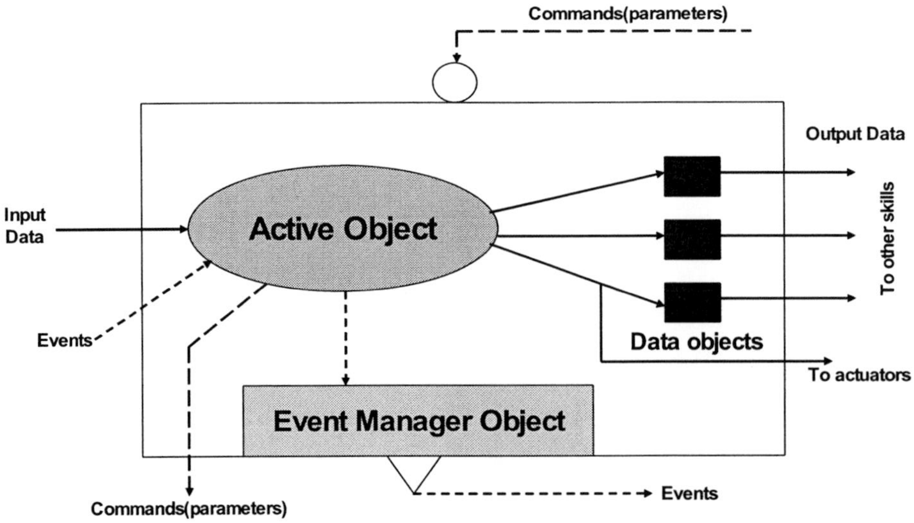

Figure 22. Skill Structure

Each skill is a distributed object that has an identity, interfaces and state. The active object of the skill has its own thread of control and it is in charge of processing. The processing results are stored in the data objects. These objects contain different data structures depending on type of data stored but the interfaces are similar. During the processing, the active object can generate events. Events are sent to the event manager object, which is in charge of notifying them to other skills, which have registered on it. In order to communicate among objects of the same module or different modules Common Object Request Broker Architecture (CORBA) is used. When a skill is activated, it connects to data objects of other skills or to sensors' servers as required by the skill. Then, it processes the received input information, and finally, it stores the output results in its data objects. If the skill is sensorimotor, it can connect to actuators' servers in order to send them movement commands. A skill can send a report about its state while it is active or when it is deactivated. For example, the skill called go-to-goal can inform on whether the robot has achieved the goal or not. When this skill is deactivated it might inform about the error between the current robot position and the goal [65].

4.3.2 User Intervention

In this model, the user has a supervisory role by which he/she can activate or deactivate the robotic skill. The error detection and recovery is perhaps the most significant challenge to time-delayed telerobotics. Errors can be handled by using a three-stage process: autonomous detection, shared diagnosis, and manual recovery. Errors are detected by using a visualization mean such as streaming video, graphical models, sensory data or connection status panels. The diagnosis task is shared by the user and the system. In recovering from the error, the user can telecollaborate with a human at the remote site or can ask the necessary privileges to be able to telnet the remote servers to reboot them.

4.3.3 Response Time

By limiting the remote interaction to high-level commands and by increasing the local autonomy of the robot using the concept of skills, the system sensitivity to the communication delay can be decreased dramatically. The total response time, which is the time elapsed between sending the skill activation command until the start of skill execution, has been obtained by measuring the time delay from client to middleware server and then from middleware server to the skill server by using software clocks. The total response time is the sum of the two round trip delays.

The results showed that the connection between the middleware and the server (radio connection 1Mbits/s) does not impose significant delay. The average value (10.89 ms) can be considered as a constant round trip delay between the middleware and the server. The major part of the delay was the client-middleware delay due to the Internet. Table 1 shows the total response time.

Table 1. Total Response Time in ms

Host	Min.	Av.	Max.
UC3M	121	138	148.65
FH-Weingarten	264.89	283.89	324.89

The measured values (138 ms from UC3M and 283.89 from Weingarten) represented acceptable values for remote laboratories applications taking into account that a maximum communication delay of 1 sec is usually taken as a reference of operability in teleoperation systems. These delays will be noticed by the users but they are accepted for an educative application as the remote laboratory.

4.4 User Interfaces

An intuitive user interface is required for inexperienced people to control the robot remotely. Beside the intuitiveness of the interface, code reuse has been considered as a design requirement. Software patterns have been used to develop the human-robot interfaces. Design patterns are design solutions [67] (not implementations) that can be used on different problems, and they have many implementations. Design patterns help software reuse, and allow the description of the structure and the collaboration scheme between components at a high abstraction level. The use of the patterns provides flexible user interfaces with minimal coupling relationships between subsystems. As shown in Figure 23, there are three types of interfaces between the user and the robot. User-machine interface or graphical user interface (GUI), machine-machine interface and machine-robot interface. The following subsections describe the patterns used to implement these three interfaces.

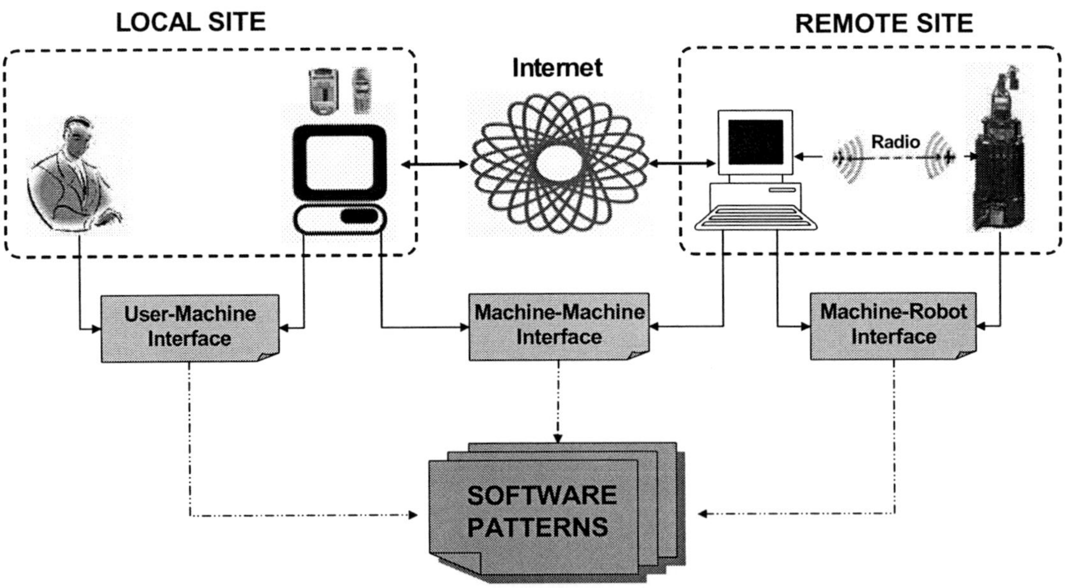

Figure 23. Patterns-based Interfaces

4.4.1 User-Machine Interfaces (Graphical User Interfaces)

The visual-proxy pattern has been used to develop system interfaces [68]. This pattern is in some ways a specialization of the Presentation/Abstraction/Control (PAC) architecture [67], which can be used to build user interfaces for object-oriented systems and can guarantee high degree of extensibility and reusability of the software components.

The objective of this pattern is to separate the generation of the user interface entirely from the abstraction layer objects to provide the reusability and extensibility facilitates [67]. As shown in Figure 24, the three objects of this pattern are the control object, the presentation object (visual proxy) and the abstraction object.

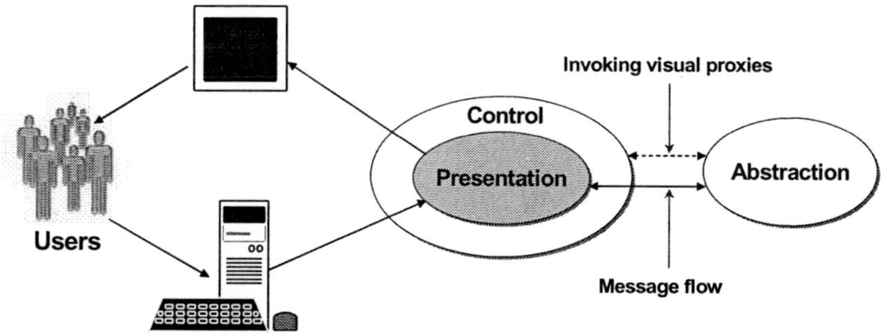

Figure 24. Visual Proxy Pattern

The control object is passive with respect to message flow. The messages go directly from the visual proxy (presentation layer) to the abstraction-layer object that manufactured the proxy. In the visual proxy architecture, the encapsulation is still intact in the sense that the implementation of the abstraction-layer object can change without the outside world knowing about it.

4.4.2 Machine-Machine Interfaces

The machine-machine communication concerns with communicate client machine with an intermediate server. The security restrictions on java applets prevent them from communicating with any machine other than the one they were retrieved from. To avoid this problem, communication was made to the server side through java servlets, which establish a communication channel between the client and the server. Servlets provide the features of traditional CGI scripts with the added benefits of efficiency and portability. Currently, major corporations are making the migration from CGI scripts to Java servlets. As a result, the demand for applet and servlet communication is on the rise. There are two mechanisms to communicate an applet front-end with intermediate server program written as Java servlet.

As shown in Figure 25, applets can communicate with servlets using the GET and POST method to send data as a string. When the applet sends data using a GET method, it must URL encode each name/value pair. Once the applet has opened a connection to the URL, then the input stream from the servlet is accessed. The applet can read this input stream and process the data accordingly.

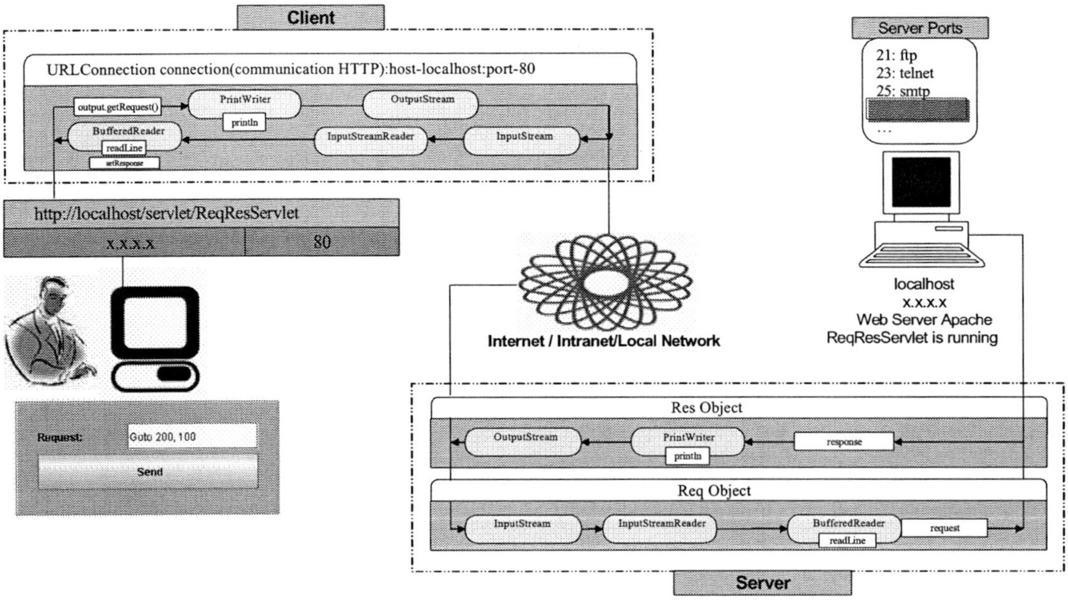

Figure 25. Client-Servlet Communication

To provide a higher level of abstraction, applet and servlet can be communicate using object serialization as shown in Figure 26. Instead of passing each parameter as name value pairs, all the parameters can be send as a true Java object. This object can be vector, array, class or file that can encapsulate all the information.

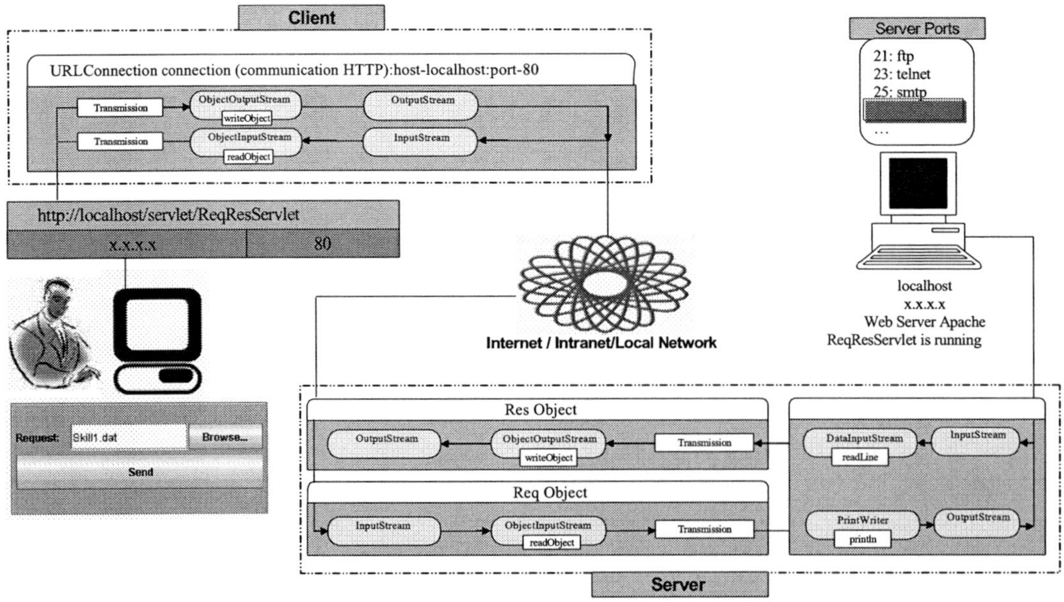

Figure 26. Object Serialization

4.4.3 Machine-Robot Interfaces

To communicate the intermediate Java server (servlets) with remote robot servers (written in C++), the broker pattern can be used. This architectural pattern can be used to structure distributed software systems with decoupled components that interact by remote service invocations. A broker component is responsible for coordinating communication, such as forwarding requests, as well as for transmitting results and exceptions [69].

The Object Request Broker of CORBA has been used in the proposed architecture as broker pattern [70]. This component provides all the communication infrastructure needed to identify and locate objects, handle connection management and deliver data and request communication. Interface Definition Language (IDL) provides a common format to represent an object that can be distributed to other applications. The other applications might not even understand objects, but as long as they can provide a mapping between the common IDL format and their own data representations, the applications can share data.

5 An Architecture for Remote Interaction

A remote laboratory for mobile robotics can be built using two kinds of distributed systems models. The first choice is to use a two-layer client/server structure, and the second option includes a three-layer model that includes a middleware. Two layer architectures are very easy to implement at first, but it is very difficult to maintain the application up to date with them. In this case, the client side is responsible for the data access, the business logic implementation, formatting the data, the user interface and data input. Because all the details involved on the client side, programs are very difficult to adapt to Internet, as they do not scale well.

As shown in Figure 27, the proposed architecture is three-layer architecture based on the concept of supervisory control. In this architecture, the client layer is responsible for the presentation of system interfaces, the middleware layer serves as controller to communicate the client with the remote robot servers and the server layer, which contains the abstraction model of the system represented by robot' servers. The following subsections describe in more details the architecture layers.

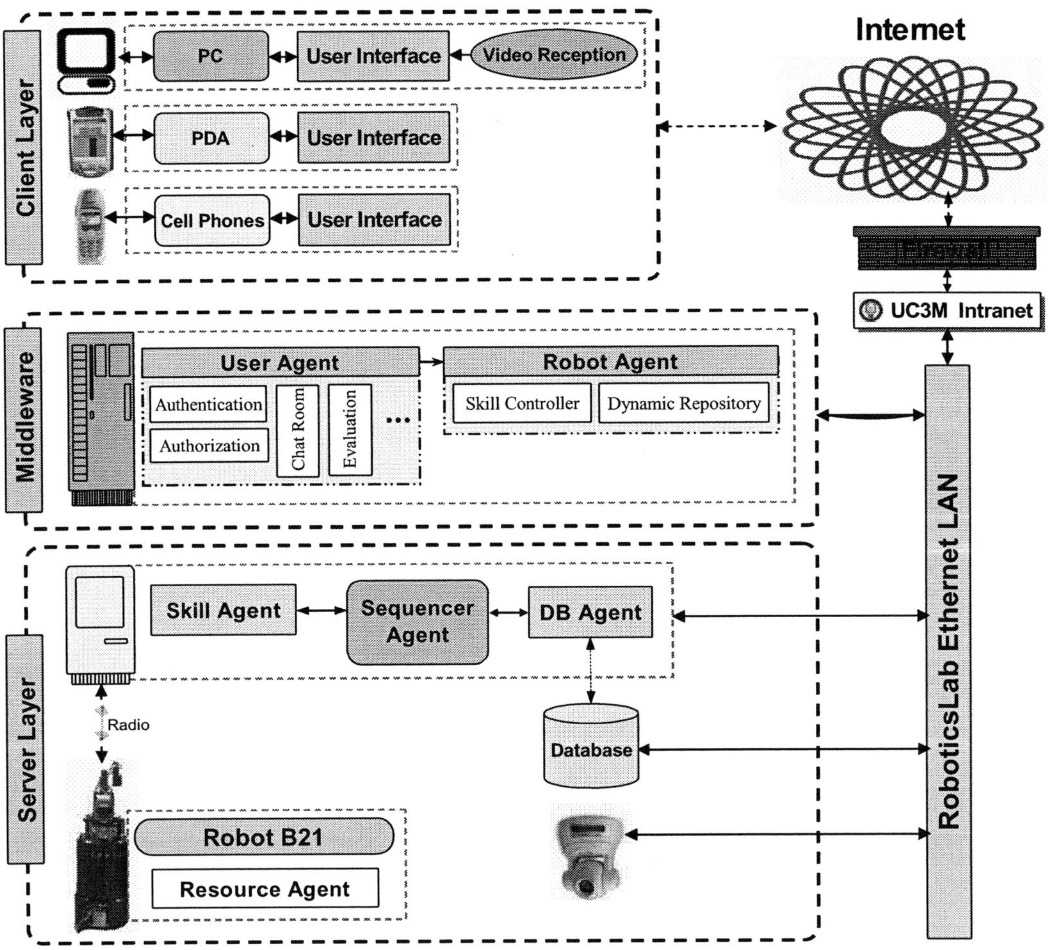

Figure 27. System Architecture

5.1 Client Layer

As shown in Figure 27, different interface agents can be developed for different interaction devices by customizing the proposed design according to the capabilities of the interaction device. The following subsections describe the client layer of the three developed controllers in the system.

5.1.1 PC-Based Controller

As shown in Figure 28, the *ExperimentTool* class implements *UserInterface* so it can produce visual proxies when asked. It asks the other *SkillController* and the *DynamicRepository* classes for visual proxies as simple *JComponents*. This class contains a constructor and implementation of all methods required by the interface but it is not a God-like class that controls everything from above. It positions the solicited proxies within the panel but does absolutely nothing else with them. This class is simply a passive vehicle for holding visual proxies. The proxies communicate directly with the abstraction-level objects that creates them and these abstraction layer objects can communicate with each other. If the state of an abstraction-level class changes as a result of some user input, it sends a message to another of the abstraction-level classes, which may or may not choose to update its own user interface (its proxy) as a sequence.

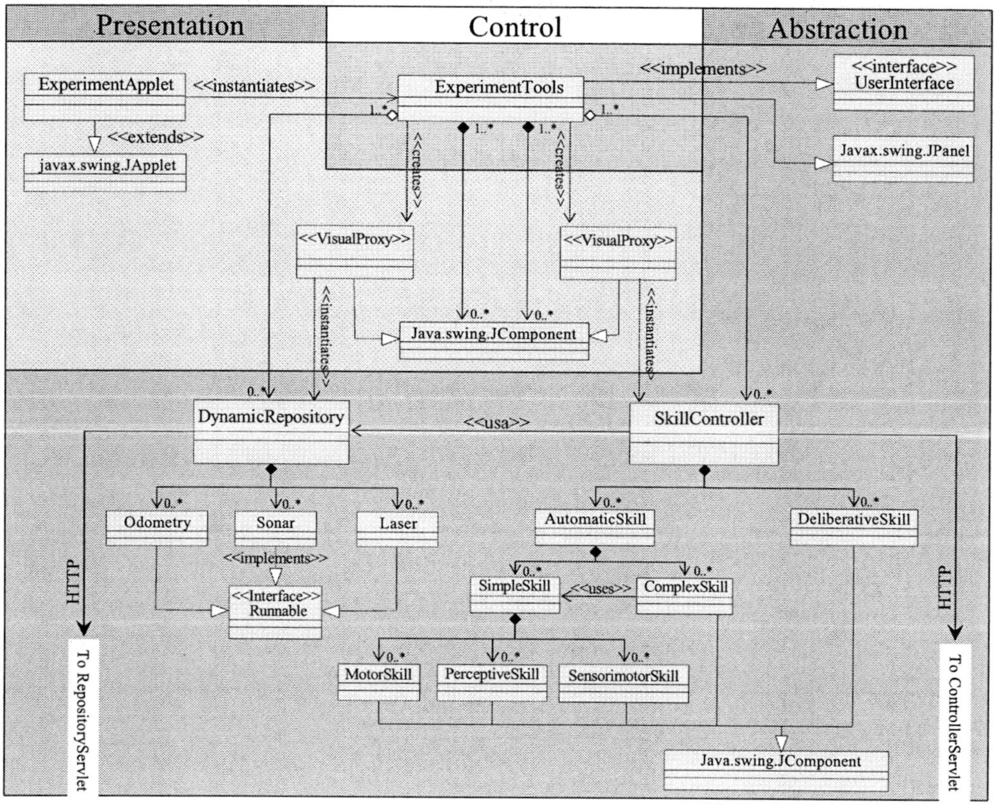

Figure 28. Client Layer

ExperimentApplet instantiates the *ExperimentTools* to form the experiment interface seen by the user via any Web browser. *SkillController* class is used to activate/deactivate the skill and it can be an automatic or a deliberative skill or a combination of different types of control classes in the same panel. The *DynamicRepository* class provides information about a sensor's status and can be used to remotely acquire the sensory data. This class is implemented as a thread by implementing the *Runnable* interface to provide updated sensory data in real time. This data may be odometry, which indicates the actual robot position and its translational and rotational velocities or ultrasonic sensor data or laser sensor data.

SkillController may use the *DynamicRepository* class to invoke sensory data, which may be necessary to complete the control task, such as in the case of obstacle avoidance skill by using sensory data. Both *SkillController* and *DynamicRepository* classes interact with the middleware using http protocol to send the user requests to the robot servers.

5.1.2 PDA-Based Controller

To develop mobile devices interfaces (PDAs and Mobile Phones), the client layer must be customized to deal with specific problems related to the limitations of handheld devices (slow processors, limited memory/storage, and small displays) and wireless network (restricted bandwidth, high latency, low connection stability, and low predictable availability).

Figure 29 shows the architecture used to develop PDA-based controller, which facilitate the remote interaction with the skill agent via skill controller agent. Compaq iPAQ 3870 PDA has been used with Compaq Wireless Client Card WL 110 for the wireless communication with the access point. An access point LINKSYS 2.4 GHz with maximum coverage rage of about 350 has been used in this experiment.

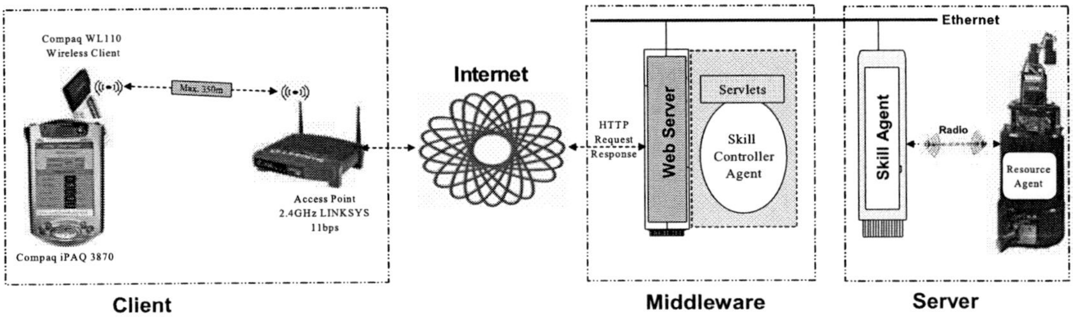

Figure 29. PDA-based Controller

The client interface has been developed using Java AWT because Java Swing is not compatible with PersonalJava 1.2 specification. A certified implementation of Sun PersonalJava 1.2 called JeodeRuntime is necessary to be installed to use the interface [15].

5.1.3 Cellular Phones-Based Controller

The Java 2 Micro Edition and the Wireless Application Protocol (WAP) are the two most common used technologies to develop wireless applications. The WAP is an open specification that addresses wireless network characteristics by adapting several data-handling approaches already in use by Web protocols, and introducing new ones, where appropriate, to the special requirements of handheld wireless devices. The reuse of existing Web technologies eases the development of WAP applications, and makes it similar to developing HTML-based Web applications since it is browser-based. Some WAP gateways are capable of converting HTML pages into a format that can be displayed on wireless devices. But because HTML was not really designed for small screens, the WAP protocol defines its own markup language, the Wireless Markup Language (WML), which adheres to the XML standard and is designed to enable powerful applications within the constraints of handheld devices [71].

The Java 2 Micro Edition (J2ME) Mobile Information Device Profile (MIDP) is also an open specification that adapts existing technologies such as Java and the Web. Developing MIDP-based applications (also known as MIDlets) is similar, but not identical, to developing Java Applets in the sense they share a similar programming model. Similar to applets, however, where an applet is described in an HTML file, a MIDlet or a group of MIDlets (known as a MIDlet Suite) is described in a Java Descriptor (JAD) file. While applets run in a Web browser, MIDlets run in a MIDlet management software (which is preinstalled on MIDP devices) that provides an operating environment for KVM and MIDlets where they run [71].

Figure 30 shows the necessary steps used to develop cellular phones-based controllers using these technologies.

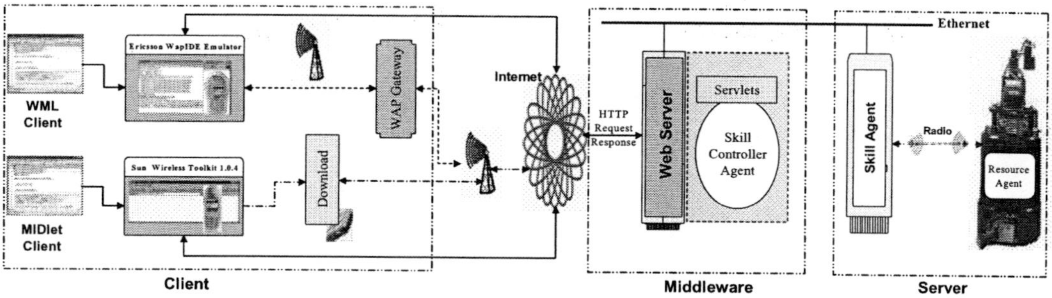

Figure 30. Cellular Phones-based Controller

In the developed interfaces, the WML-based interface consists of multiple cards in a deck whereas the J2ME-based interface consists of several screens, but only one screen at a time can be visible. Figure 31 shows a comparison between the view of a WAP application and MIDlet.

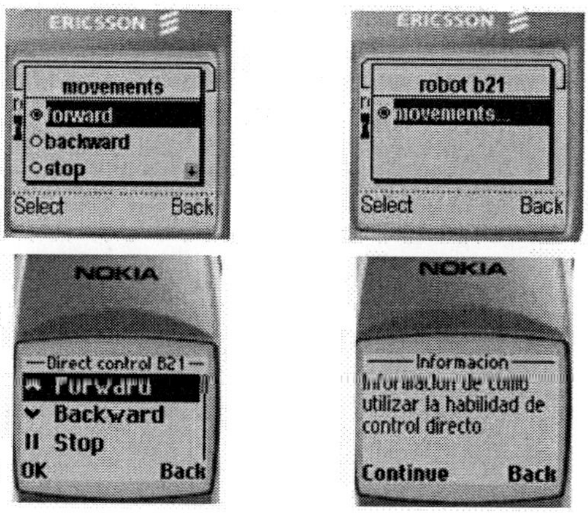

Figure 31. WML-based Interface

Many interfaces have been developed to send direct control commands to move the robot B21 forward, backward or to turn it left or right and to activate or deactivate various movements skills such as rotate, wall following, round trip and gotogoal. The conducted experiments through this work have shown that these technologies improve the reachability and the usability of remote interaction environments.

5.2 Middleware Layer

In this layer, there is a PC linked internally to an Ethernet LAN and externally to the university's Intranet which runs an Apache Web server, which hosts two agents:

5.2.1 Robot Agent

This agent deals with the real robot when the robot is actually running. It provides the required information to process a certain skill selected by the user or required in step of an experiment. The agent contains two groups of java servlets, dynamic repository servlets to invoke the sensory data from a resource agent in the server layer and skill controller servlets that send control commands to the skill agent in the server layer. The communication between these servlets and the remote robot servers is done via the Object Request Broker (ORB) of the Common Object Request Broker Architecture (CORBA) where the Java servlet acts as a client to the robot server. The ORB provides the communication via the unified interface language Interface Definition Language (IDL) and based on the Internet Interoperable ORB protocol (IIOP). The decision to use CORBA as the distributed object architecture of the remote laboratory is based on a qualitative and quantitative comparison between the two most commonly used architectures, CORBA and RMI [70]. This study concluded that CORBA is suitable for large scale or partially Web-enabled applications where legacy support is needed and good performance under heavy client load is expected. Moreover, CORBA servers can be located at any Internet site. RMI, on the other hand, is suitable for small-scale fully Web-enabled applications where legacy support can be managed with custom build or pre-built bridges, where ease of learning and ease of use is more critical than performance. Integrating Java RMI with IIOP can provide another solution that combines RMI easy programming features with CORBA interoperability [72].

5.2.2 User Agent

This agent contains two main servlets to manage user access. The user authentication servlet collects and verifies the user registry. The authorization servlet manages the authenticated user access and notifies the user that the access to the robot agent is granted or rejected based on round robin approach. When only a single user is connected, no restrictions on usage are given. However, if multiple users are simultaneously connected, only one is given access for a specified amount of time. Once the time has expired, the next user in line will be given opportunity to control the robot and the first user will be moved to the end of the waiting list. A chat room is provided as a communication mechanism between the different users. Also this agent holds other servlets necessary for the teleeducation environment as simulation servlet, system evaluation servlet, online test corrector servlet, etc. These services do not require connection with the real robot and they are accessible to all connected users.

5.3 Server Layer

This layer contains four main agents. Resource agent represents the servers of robot's resources implemented in C++ such as base server (actuator server, odometry server and sonar server), laser server, etc. This agent is also responsible for managing the users' requests to avoid any data inconsistency. The skill agent represents skill server, which may be simple automatic skill or deliberative skill server. The simple skills can be combined to form complex skill by using the sequencer agent [73]. The sequencer is responsible for deciding which skills have to be activated in each moment to avoid simultaneous activation of skills which act upon the same actuator. The data base agent provides access to a database, which contains plans of sequences. This layer also contains a commercial network camera (Sony

SNC-RZ30P) with embedded streaming video and pan tilt zoom control capabilities to provide real imagery transmission from the remote environment to the local site of the operator.

5.4 Interaction between Layers

As shown in Figure 32 to activate or deactivate a robotic skill, the user has to use the SkillGUI to send Http-based commands to the SkillServer through SkillServlet, which servers as an intermediate server between client and server.

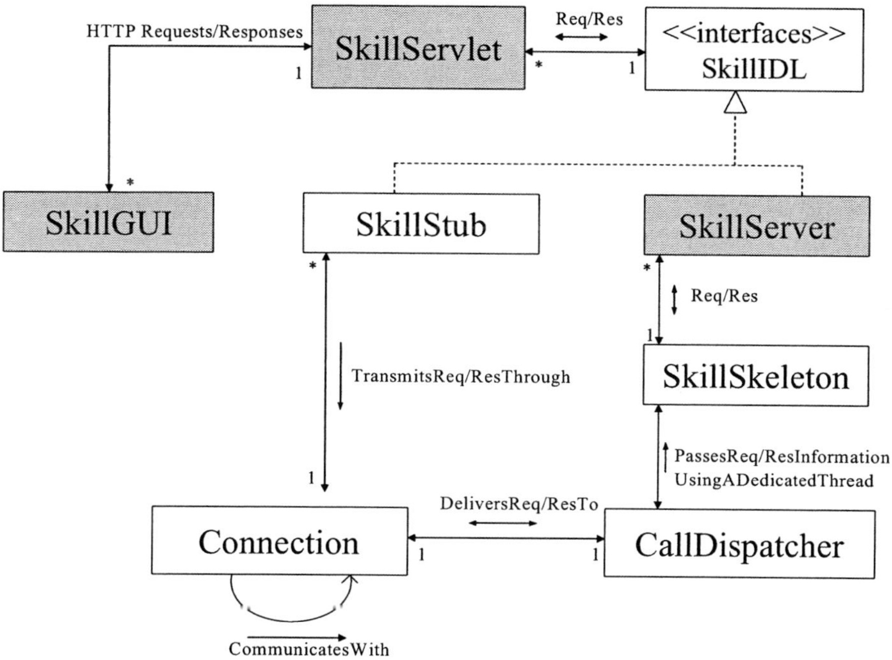

Figure 32. Client/Server Interaction

SkillServlet does not call the SkillServer directly but through an Interface Definition Language-IDL interface (SkillIDL). This class calls methods of a proxy object that implements the SkillIDL interface. Because it calls methods through an interface, it doesn't need to be aware of the fact that it is calling the methods of a stub object that is a proxy for the SkillServer object rather than the SkillServer object itself.

The stub object encapsulates the details of how calls to the SkillServer object are made and its location. These details are transparent to SkillServlet objects. Also stub objects assemble information identifying the SkillServer object, the method being called and the values of the arguments into a message. On the other end of the connection, part of the message is interpreted by a CallDispatcher object and the rest by a skeleton object. The connection classes are responsible for the transport of message between the environment of a remote SkillServlet (Web Server) and the environment of the SkillServlet (Robot Server).

The CallDispatcher receives message through the connection object from a remote stub object and then passes each message to an instance of an appropriate skeleton class, which are created by the CallDispatcher. The CallDispatcher object are responsible for identifying the SkillServer object whose method will be called by the skeleton object.

The skeleton classes are responsible for calling the methods of SkillServer objects on behalf of remote SkillServlet objects. The skeleton object extracts the argument values from the message passed to it by the CallDispatcher object and then calls the indicated method passing it the given argument values. If the called message returns a response, then the skeleton object is responsible for creating a message that contains the return value and sending it back to the stub so that the stub object return it. If the called message throws an exception, the skeleton object is responsible for creating an appropriate message.

The SkillServer classes (implemented in C++) implement SkillIDL interface. Instances of this class can be called locally through the SkillIDL interface, or remotely through a stub object that also implements the same skill interface.

The following steps can be used to implement skill client/server interaction using the http request/response protocol and the broker pattern as shown in Figure 33.

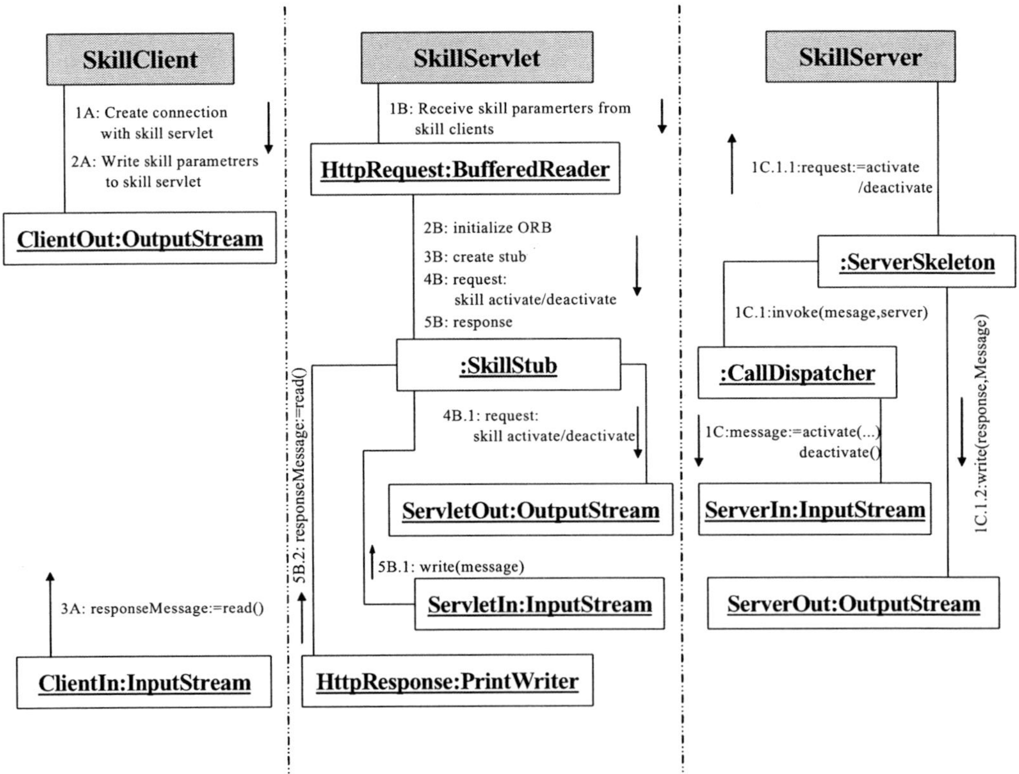

Figure 33. Interaction Steps

1A: The SkillClient object creates an HTTP-based connection with the skill servlet.

```
URL SkillServletURL = new URL(url);
URLConnection SkillServletConnection =
SkillServletURL.openConnection();
SkillServletConnection.setDoOutput(true);
SkillServletConnection.setUseCaches(false);
```

2A: The client sends to the servlet the necessary parameters to activate/deactivate the skill.

```
PrintStream out = new
PrintStream(SkillServletConnection.getOutputStream());
out.println(param1);
out.println(param2);
out.println(param3);
.
.
.
out.close();
```

1B: SkillServlet receives the parameters sent by the client to activate/deactivate the skill.

2B: A call will have been made by the servlet to initialize the object request broker (ORB).

```
Properties props = new Properties();
props.put("org.omg.CORBA.ORBInitialPort",#);
String[] args = new String[]{""};
ORB orb = ORB.init(args, props);
```

3B: The servlet obtains the stub object to call the remote server object. ORBs provide a mechanism that takes the logical name of an object and creates a stub object in consultation with a mechanism that uses the Registry pattern to provide location of the server object.

```
org.omg.CORBA.Object objRef =
orb.resolve_initial_references("NameService");
NamingContext ncRef = NamingContextHelper.narrow(objRef);
NameComponent nc = new NameComponent("Skill", "Framework");
NameComponent path[] = {nc};
org.omg.CORBA.Object objRefSkill = ncRef.resolve(path);
FrameworkComponents.SystemModule SkillRef =
FrameworkComponents.SystemModuleHelper.narrow(objRefSkill);
FrameworkCore.ObjectContainer container = SkillRef;
String[] skill_i = new String[]{"Skill_i"};
org.omg.CORBA.Object obj = container.find_object(skill_i);
SkillRef = RobotSkill.SkillHelper.narrow(obj);
```

4B: The servlet calls the SkillStub object to activate/deactivate the skill with the intention of calling the skill object that may or may not be remote.

```
SkilltRef.activate (param1, param2, param3, …);
```

4B.1: The SkillStub object asks the ServletOut object to write a message that includes the skill class name, the method name and the arguments that the servlet object passed to activate/deactivate the skill. The ServletOut object passes the message through a network connection to the ServerIn object.

1C: The CallDispatcher object extracts the class of the object to call from the message. It then obtains the actual object whose method is to be called. Using different thread, the

CallDispatcher object asynchronously calls the invoke method of the SkillSkeleton object. It passes the message and the server object whose method is to be called to the invoke method.

1.C.1.1: The SkillSkeleton object extracts the name of the method to call and the arguments from the message. It then calls the server objects's activate/deactivate method.

1.C.1.2: The SkillSkeleton object constructs the response produced by the call to the skill object's activate/deactivate method. This can be a returned value or a thrown exception. It then passes the message to the ServerOut object's write method, which passes the message through a network connection to the ServerOut object.

5B: The servlet returns the response message to the client object.

5B.1: The ServletIn object's read method returns the response message to the SkillStub object.

5B.2: The SkillStub object extracted the response from the message. If the response is a value, it returns it in form of HTTP response. If the response is an exception, it throws it.

3A: The ClientIn object reads the response through a network connection (Internet) and returns it to the SkillClient object.

6 Case Study – Innoavtive Educational Environment for Mobile Robotics

Robotics challenges traditional engineering disciplines because the integrated approach in the selection of means to the intended functional ends must involve, by nature, a team activity, and crossing the boundaries between conventional engineering disciplines. Unlike traditional fields, robotics is still an emerging area that combines the essential elements of mechanical engineering, electrical engineering, and computer science. Relatively few robotics education programs exist at the graduate level, and even fewer exist at the undergraduate level. The courses in existence are still new and are open to rapid change and new approaches. Course goals can change from year to year as new technologies and theories are introduced into the field at large.

To build an educational environment for a field like mobile robotics, it is recommended that one combines the traditional educational tools with telematics-based tools. The telematics-based tools can support the traditional tools and can solve the traditional teaching constraints of curriculum, class size and limited resources. However, these tools cannot totally replace traditional education tools, especially in a cross-disciplinary field like robotics. A real interaction with a robot is necessary to give the student hands-on robotics experience and to help them to understand real-world problems and be able to identify these problems and formulate cost-effective applicable solutions.

We proposed an approach to build innovative environments for mobile robotics education [74]. The proposed approach can be used to create new courses or to update courses, which are currently taught by traditional way. In this approach, different educational activities are combined in order to develop an innovative teaching environment in the field of indoors mobile robotics (Figure 34).

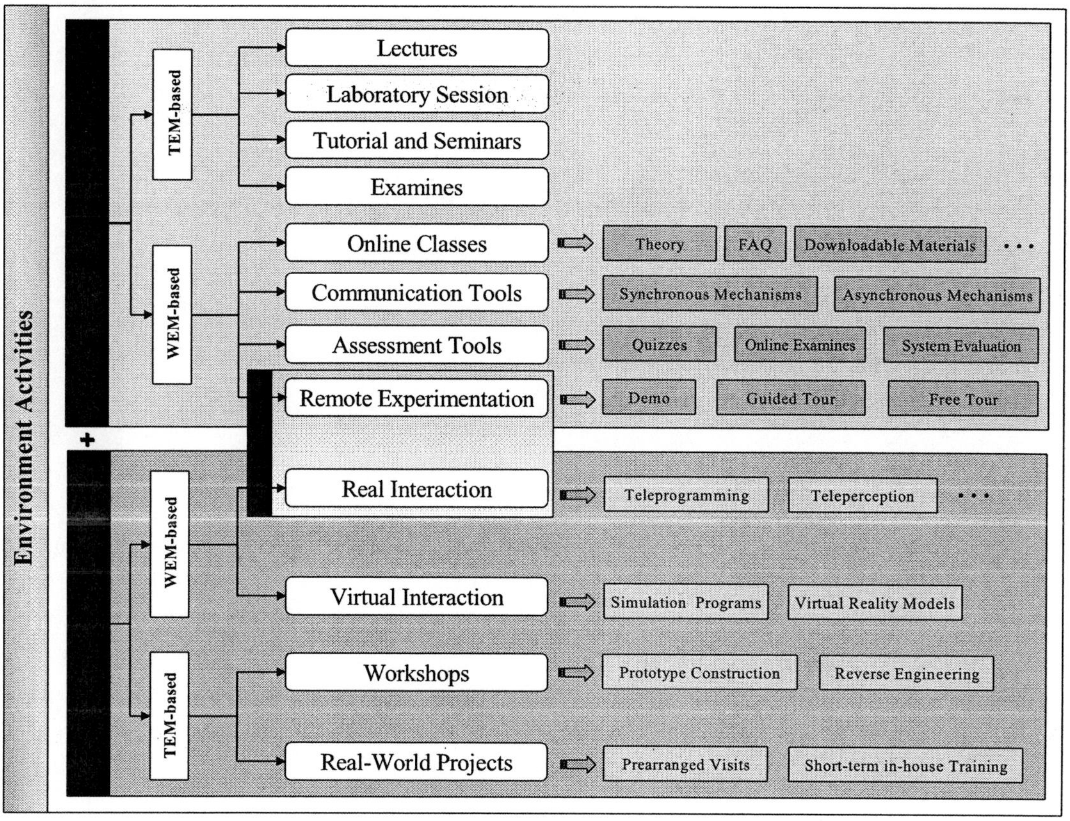

Figure 34. Environment Activities

These activities are classified into instructional activities and constructional activities. The instructional activities aim at presenting theoretical and practical knowledge of the course to the students. The constructional activities give the students the opportunity to physically construct and implement the ideas derived from the course.

The instructional and constructional activities have been implemented using both Web-based education model (WEB) and traditional education model (TEM). The Web-based tools have been used to support traditional tools and to increase the feeling of control that students have over their learning experience, which is one way to increase student motivation. Nevertheless these tools cannot totally replace traditional educational tools especially in a cross-disciplinary field like robotics. As mentioned previously, a real interaction with a robot is necessary.

The proposed approach has successfully been used to build an innovative educational environment in the field of indoor mobile robotics at Carlos III University of Madrid. This environment is not a public access environment and only IECAT partners and Carlos III University community members can use it. This section describes the using of this approach to update a postgraduate course on intelligent autonomous vehicles.

6.1 Intelligent Autonomous Vehicles Course

This course is offered to postgraduate students within an integrated interdepartmental doctoral program of electrical, electronic and automatic engineering. This course offers different theoretical and experimental knowledge about intelligent autonomous robots. Based on the proposed approach many activities have been integrated to update the teaching strategy of this course. The following subsections described the instructional and constructional activities, which have been implemented and integrated into the course modules in order to update them.

6.1.1 Instructional Activities

Many instructional activities have been implemented based on a web-based education model to support the traditional teaching.

6.1.1.1 Online Classes

Many online classes have been designed to present the basic knowledge of robots as a general issue and indoor mobile robotics as main subject. This knowledge includes introductory concepts, robots classifications, mobile robot anatomy, control architectures of mobile robots, sensors commonly used in mobile robots and mobile robots applications. The class also has been supplemented by FAQ pages, a search engine, a digital library and downloadable materials.

6.1.1.2 Remote Laboratory

An online laboratory in a field such as mobile robotics must have live performance characteristic, not just virtual reality or simulation programs. The multi-layered architecture described previously has been implemented in order to reach this goal. The knowledge is presented to the user by using three types of tutoring tours. These tours are classified according to the level of guidance of the tutor into: fully guided, unguided, and guided tour. The fully guided tour is a demonstration tour, which demonstrates basic concepts without any intervention on the part of the student. The unguided tour or free tour does not determine any order and tasks at all. This tour provides generic tools to the user and let him/her customize the experiment according to his/her needs. These generic tools include 2D model for the robot and the lab, odometry data panel, sonar data panel, laser data panel, motion controller and network camera. The guided tour aims to present different specified courses with direct interaction between the student and the tutor. The experimental issues of the courses are provided by using interfaces of different types of skills as shown in Figure 35.

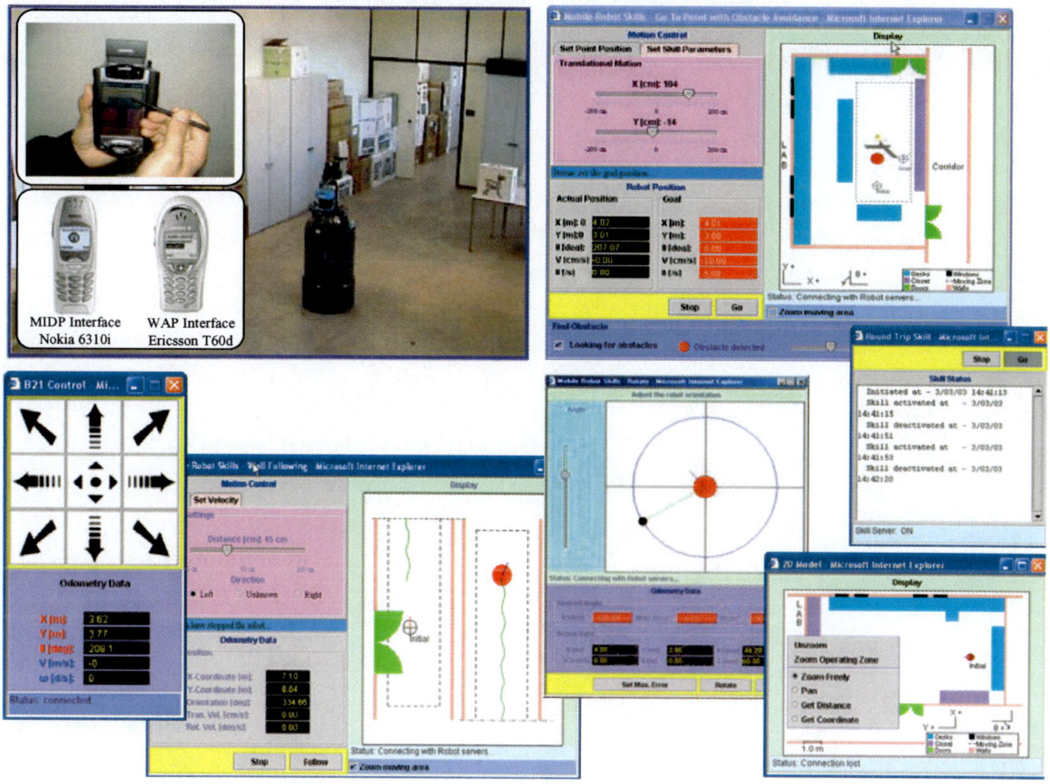

Figure 35. Screenshot of skill Interfaces

The following examples are some of the experiments that have been implemented and are presented through the guided tour.

Direct Control: This experiment aims at familiarizing the user with the mobile robot motion control and positioning by using different interaction elements such as PC or any MIDP (Mobile Information Device Profile)-enabled device such as handheld PDA or cellular phones. The remote user can send direct control commands to move the robot B21 from RWI forward, backward or to turn it left or right. Using a network camera, a 2D graphical model, and the odometry data (actual location with respect to the initial point, translational and rotational robot velocities), the remote user will be able to view the effect of the sent commands.

Orientation Control Skill: This skill provides a closed loop control on the robot orientation. The operator determines the desired orientation and the maximum error and then the skill can be activated to change the actual robot orientation to the desired one.

Go-to-Goal Skill: By using this automatic skill, the user can send the robot to a certain point in the lab. To activate this skill the user has to set the skill parameters as goal point, robot velocity and maximum error. The skill can provide information as to whether the robot has achieved the goal or not. When this skill is deactivated it might supply information about the error between the current robot position and the goal. This skill also has been combined with a perceptive skill called detect obstacle to avoid the obstacle during motion.

Wall Following: By this skill, the robot can follow certain contour using sonar data. The distance between the robot and the contour and the robot's velocity are controlled by the user.

Round Trip Skill: This skill is a complex skill formed by combining go-to-goal skill and orientation control skill. It commands the robot to travel to a certain position and then return to the initial point. This skill can be used to calculate the systematic odometry errors based on University of Michigan Benchmark test (UMBmark) [75].

Environment Modelling and Robot Localization: The modelling skill can be used to construct an environment map using sensory data. The constructed map is then used to remotely estimate the robot's position by computing sets of poses which provide a maximal-quality match between a set of current sensor data and the map. The remote user can compare the result of the localization algorithm with the odometry data to determine the error as shown in Figure 36.

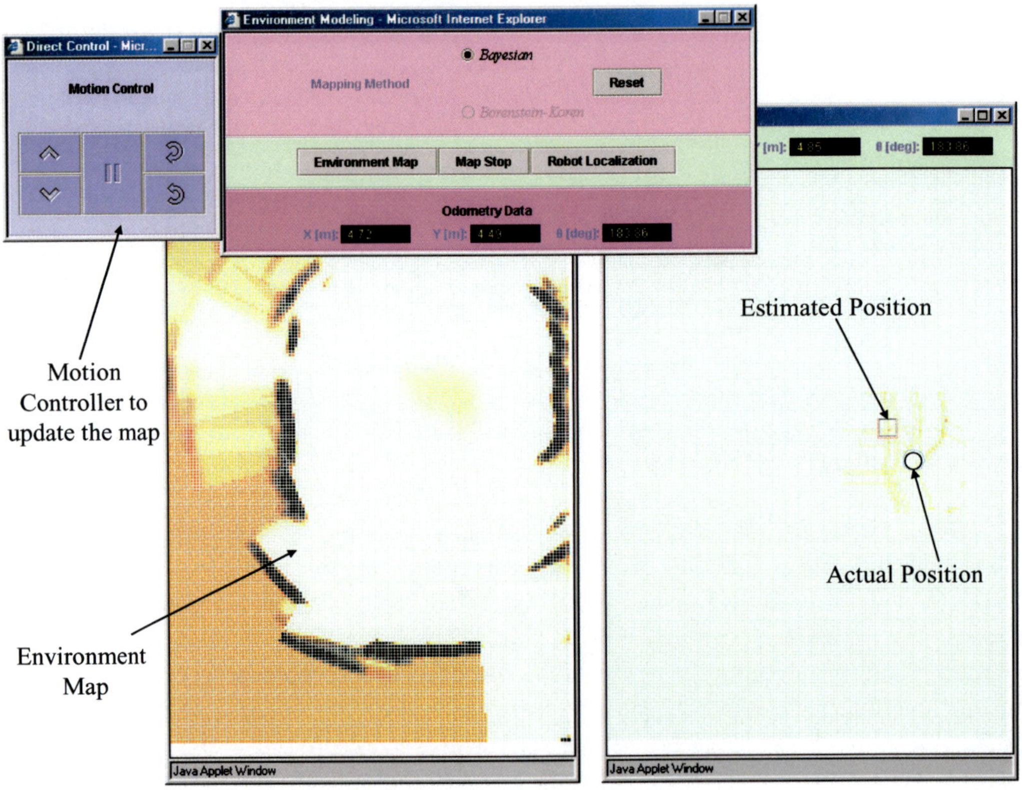

Figure 36. Modelling & Localization Experiment

6.1.1.3 Assessment Tools

Evaluative tests can be used to provide online assessments. A student first takes an exam and then, based on the results of the exam, customized learning activities can be generated to reinforce the areas in which the student scored poorly. Moreover practice tests can be designed to cover the experimental issues as a simulation to traditional laboratories. Three types of tests have been designed as online assessment tools. Online quizzes are used to evaluate the student's background in mobile robotics in general. Time restricted theoretical tests are used to evaluate the student's knowledge in the theoretical topics of the course. Time restricted experimental tests are designed to help the student to understand practical problems.

6.1.2 Constructional Activities

Two types of constructional activities have been implemented based on the two education models, Web-based education model (WEB) and traditional education model (TEM).

6.1.2.1 TEM-Based Activities

TEM-based activities such as the design and the construction of new prototypes or reverse-engineering of existed systems are used to provide face-to-face interaction between the students and teaching staff and between the student and the machine, and therefore increase student motivation and decrease students' feeling of isolation. These constructional activities also help to increase student creativity and team work. Students are asked to build robotic prototypes using lab kits such as Vex Robotics Starter Kit, Lego or Fischertechnik parts (including motors) as hardware components, the handy board (Motorola 88HC11 based or 16-bit Siemens 80C167) micro controller board for control, commonly used sensors and other components. A teleoperated rover also can be built and programmed by the students. Figure 37 shows prototype construction process in the Robotics Lab.

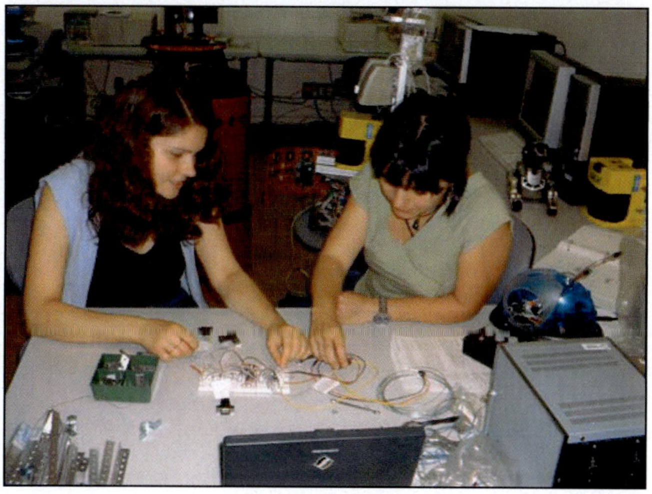

Figure 37. Prototypes Construction

6.1.2.2 WEM-based Activities

The WEM-based constructional activities include interactive learning activities as teleprogramming, telemonitoring, telediagnostic, telemaintenance and teleperception, etc., which can be incorporated into the remote laboratory to provide real interaction with the robot. The following subsections described two of implemented experiments to cover these activities.

Teleperception: The objective of this experiment is the perception of environment using multi-sensor data (sonar and laser). The experiment is divided into two parts: without robot motion and with robot motion. The objective of the first part is to understand the operation of sonar and laser sensors and to be familiar with these readings. The second part aims at recognizing the real environment using sensor data. Figure 38 shows the use of accumulative readings of sonar and laser sensors in determining the obstacle zone remotely.

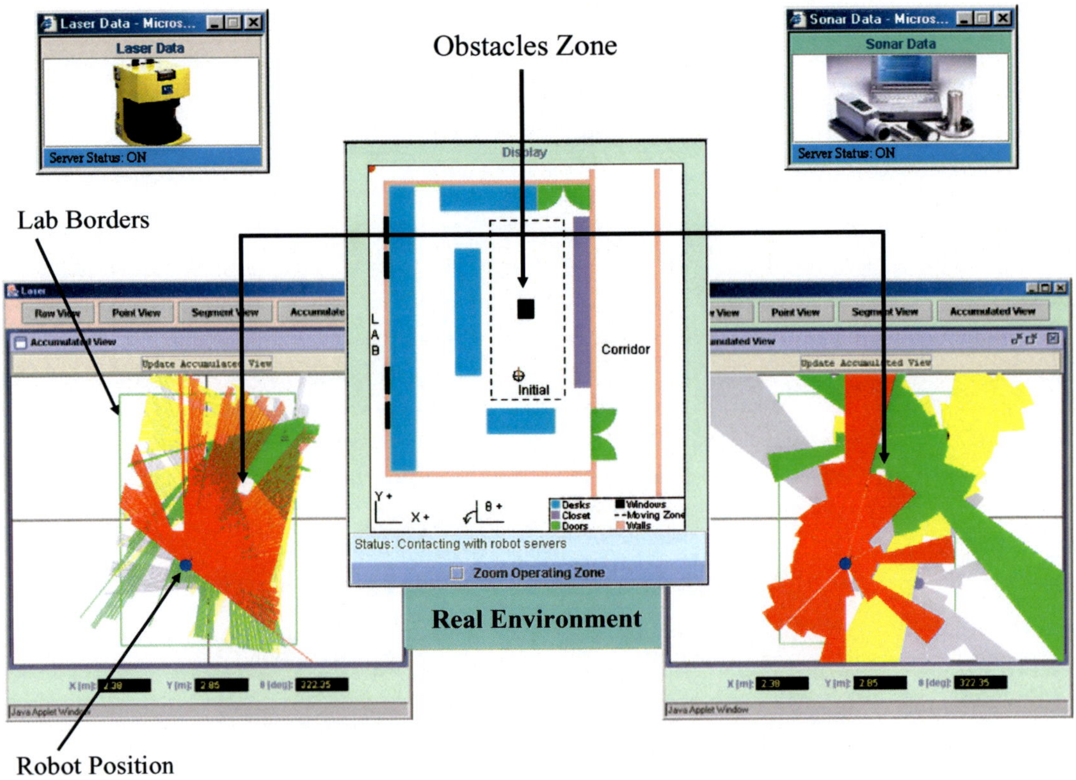

Figure 38. Environment Perception Experiment

Teleprogramming: Teleprogamming provides an asynchronous approach, which helps student to create a series of commands that is then downloaded into the robot and executed in order to examine new algorithms or realize a certain mission. An interface has been developed to help students to write a sequence plan to combine simple skills in order to obtain more complex skills. Figure 39 shows a screenshot of the sequencer interface and the sequence plan of GoToLab complex skill. This skill is used to send the robot to a certain room and it is generated using the sequencer by combining varies simple skills such as GoToPoint, FindDoorLaser, ApproachDoor, CrossDoor and WallFollowing [73].

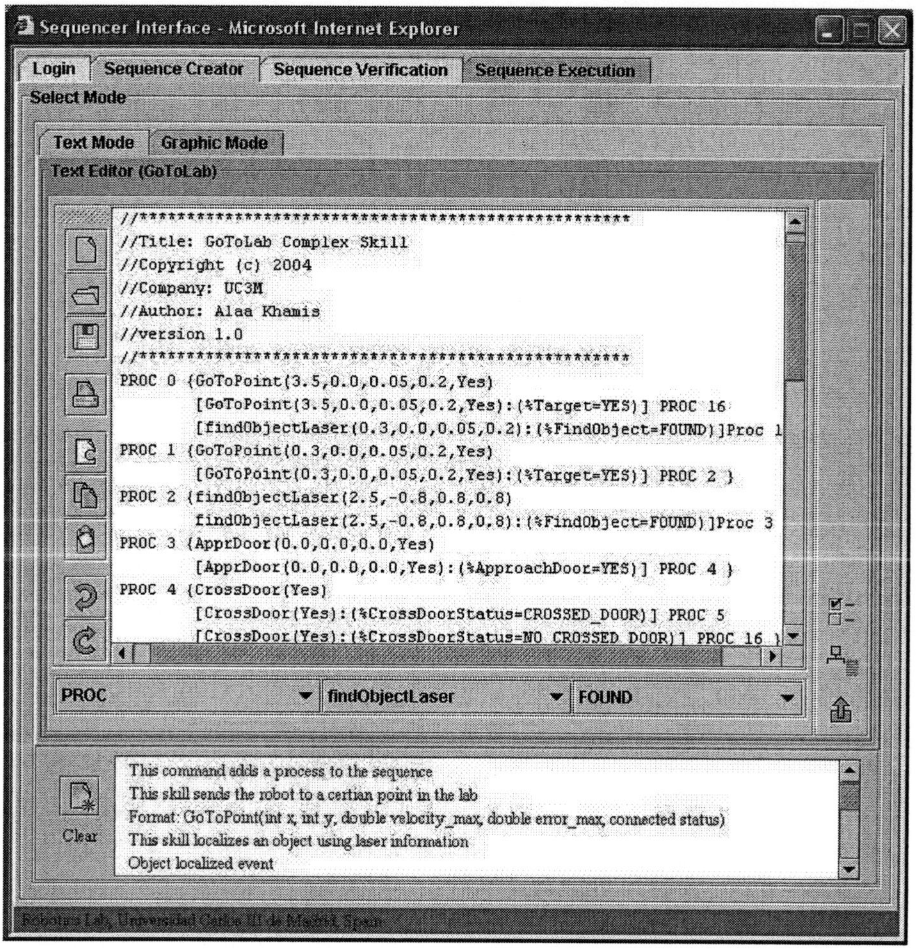

Figure 39. Teleprogramming

In addition, a low-level commands editor is now under development to facilitate interacting with the remote robot using low level commands or a group of commands in the form of a programming script. The user will be able to teleprogram the robot or to send low level commands via Internet to the robot's base server to perform many tasks such as adjusting the robot control parameters, increasing/decreasing robot velocity, informing about battery state, setting the Watch Dog timer, which can be used for safety or diagnostic purposes, etc.

6.2 System Evaluation

Student feedback was gathered using an online questionnaire. The student responses were uniformly positive as to the use of the different proposed teaching activities specially the use of the remote laboratory. Most of the students felt that the online experiments helped them to achieve a deeper understanding of the subject material. Table 2 shows examples of students' feedback.

Table 2. Student Feedback

Question	Agree	Strongly Agree
The learning objectives were clear	43%	57%
The course was well presented	29%	71%
The lab was easily accessible	29%	71%
The corrected exercises contained helpful comments	43%	57%
The system was essential to the course	14%	86%
The system was reliable	43%	57%
The system was easy to use	29%	71%

7 Conclusion

This chapter described the development of software architecture for building a remote laboratory in the field of mobile robotics to facilitate the remote interaction with mobile robots using varies interaction devices. This remote laboratory has been developed as a part of the IECAT project, which aims at developing a network of remote laboratories between six universities in Europe and United States in the field of autonomous and teleoperated systems. This international distributed laboratory provides a coordinated set of experiments for students with hardware facilities physically spread over different locations, but accessible via the Internet. It permits the exchange of existing hardware resources and educational materials between the partners.

In this chapter, the advantages and disadvantages of using the Internet as a communication medium in remote interaction systems with mobile robot are studied. A survey was conducted to evaluate the network performance between the University Carlos III in Madrid (UC3M), and the University of Applied Sciences (FH-Weingarten) in Germany, which are partners in the IECAT project. This study has focused on the most common problems of the Internet such as time delay, bandwidth and jitter.

Supervisory control paradigm has been used to develop the proposed system in order to alleviate high communication data rates and system sensitivity to network delays.

In the chapter, the use of software patterns is proposed for developing flexible and reusable human-robot interfaces. The user interfaces have been developed using visual proxy pattern. This pattern provides flexible user interfaces with minimal coupling relationships between subsystems. The generation of the user interface is entirely separated from the abstraction layer objects to provide the reusability and extensibility facilitates.

The use of mobile devices such as PDAs and cellular phones as interaction devices with the robot is studied. The conducted experiments through this work have shown that these technologies improve the reachability and the functionality of remote interaction environments.

Finally, an approach for building innovative educational environments for mobile robotics is proposed. This approach combines varies activities to solve the problems of the traditional paradigm of education and to permit the students to have an active and creative role in the system.

References

[1] P. Backes, K. Tso, J. Norris, G. Tharp, J. Slostad & R. Bonitz, "Mars Polar Lander Mission Distributed Operations", *IEEE Aerospace Conference*, 2000, Big Sky, USA.

[2] P. Backes, K. Tso, J. Norris, G. Tharp, J. Slostad, R. Bonitz & K. Ali, "Internet-Based Operations for the Mars Polar Lander Mission", 2000 *IEEE Intl. Conf. on Robotics and Automation*, ICRA00, San Francisco, USA.

[3] RedRover, Available at: http://redrover.ars.fh-weingarten.de/

[4] G. Thomas, M. Fienup, P. Gray, N. Changnon, P. Debbins, R. Dixon, T. Ferris, L. Hirshy, H. Jongedyk, A. Locke, D. Perry, M. Symonds, and R. Tahir. *Human Control of Cooperative Robots Working on Mars*. Technical report, Iowa Space Grant Consortium, 2001.

[5] S. Venema and B. Hannaford, "Telerobotic Remote Handling of Protein Crystals via an Internet Link", In K. Goldberg and R. Siegwart (Editors), *Beyond Webcams: An Introduction to Online Robots*, Cambridge, MA: MIT Press, 2002.

[6] K. Nakamura, "Companion' robot represents the future in nursing home care", News 24 Houston, October 28th 2003.

[7] J. Songmin, H. Yoshiro, Y. Gang, T. Kunikatsu, "Distributed Telecare Robotic Systems Using CORBA as a Communication Architecture", *The 2002 IEEE International Conference on Robotics and Automation*, ICRA 2002, pp.2202-2208.

[8] W. Siu, A. Ferworn, and K. Plataniotis, "Constrained image understanding for an internet robot supporting telepresence," *Proceedings of IEEE Canadian Conference on Electrical and Computer Engineering*, 2001.

[9] S. Jia, Y. Hada, and K. Takase, "Distributed Telerobotics System based on Common Object Request Broker Architecture," *Journal of Intelligent and Robotic Systems*, vol. 39, pp. 89–103, 2004.

[10] M. Stein, "PumaPaint Project", *Journal of Autonomous Robots* **15**, pp. 255–265, 2003, Kluwer Academic Publishers.

[11] S. Goldberg, "Digimuse: An Interactive Telerobotic System for Remote Viewing of 3D Art Objects," *IROS98*: *Workshop on Web Robots*, pp. 55–60, 1998.

[12] K. Goldberg, S. Gentner, C. Sutter & J. Wiegley, "The Mercury Project: a Feasibility Study for Internet Robots", IEEE Robotics and Automation Magazine, 7(1):35-40, 2000.

[13] P. Saucy & Mondada, "KhepOnTheWeb: Open Access to a Mobile Robot on the Internet", IEEE Robotics and Automation Magazine, 7(1):41-47, 2000.

[14] J. Fernandez, P. Lopez, and J. Oliva, "A Distributed Multirobot System based on Edutainment Robots," Proceedings 2005 IEEE International Conference on Robotics and Automation, vol. 2205, pp. 4260 – 4265, 2005.

[15] Insignia Solutions, http://www.insignia.com/

[16] T. Fong, F. Conti, S. Grange, y C. Baur, "Novel Interfaces for Remote Driving: Gesture, Háptic and PDA", SPIE 4195-33, SPIE Telemanipulator and Telepresence Technolgies VII, Boston, MA, 2000.

[17] M. Salichs, F. Rodriguez, D.M. Rivero, A. Khamis, "Pattern-oriented Implementation for Automatic and Deliberative Skills of a Mobile Robot." The 1st International Workshop on Advances in Service Robotics, ASER03, March, 13-15, 2003, Italy.

[18] R. Simmons, R. Goodwin, K. Haigh, S. Koenig, and J. Sullivan, "A layered architecture for office delivery robots," *International Conference on Autonomous Agents* (Agents'97), pp. 235–242, 1997.

[19] R. Simmons, R. Goodwin, K. Haigh, S. Koenig, J. O'Sullivan, and M. Veloso, "Xavier: Experience with a Layered Robot Architecture," *SIGART Bull.*, vol. 8, pp. 22–33, 1997.

[20] T. Fong, C. Thorpe, and C. Baur, "Multi-robot remote driving with collaborative control", *IEEE Transactions on Industrial Electronics*, 2003.

[21] L. Ngai, W. Newman, and V. Liberatore, "An Experiment in Internet-based, Human-assisted Robotics", *ICRA*, pp. 2190–2195, 2002.

[22] G. Caprari and R. Siegwart, "Design and control of the mobile micro robot Alice", *Proceedings of the 2nd International Symposium on Autonomous Minirobots for Research and Edutainment*, pp. 23–32, 2003.

[23] D. Zhang & L. Wang, "Web-based remote manipulation in advanced manufacturing", *IEEE Int. Workshop on Business Service Network*, 2005.

[24] H. Hu, P. Tsui, L. Cragg & N. Volker, "Agent architecture for multi-robot cooperation over the internet", International Journal of Integrated Computer-Aided Engineering, vol. 11, pp. 213–226, 2004.

[25] A.Weitzenfeld, S. Gutierrez-Nolasco & N. Venkatasubramanian, Controlling Mobile Robots with Distributed Neuro-Biological Systems*", The 2nd Annual Symposium on Autonomous Intelligent Networks and Systems (AINS'03),* 2003.

[26] Decommissioning of Nuclear Installations, "Decommissioning in short," http://www.eu-decom.be/about/decominshort/whatis.html, 2006

[27] L. Cragg and H. Hu, "Application of mobile agents to robust teleoperation of internet robots in nuclear decommisioning*", Proceedings of IEEE International Conference on Industrial Technology - ICIT'03*, pp. 10–12, 2003.

[28] L. Parker, "Alliance: an architecture for fault tolerant multirobot cooperation", *IEEE Transactions on Robotics and Automation*, vol. 14, pp. 220–240, 1998.

[29] K. Goldberg, B. Chen, R. Solomon, S. Bui, B. Farzin, J. Heitler, D. Poon, and G. Smith, "Collaborative teleoperation via the internet*", IEEE International Conference on Robotics and Automation*, 2000.

[30] K. Goldberg, D. Song, and A. Levandowski, "Collaborative teleoperation using networked spatial dynamic voting", *IEEE, Special issue on Networked Robots*, vol. 91, pp. 430–439, 2003.

[31] D. Song and K. Goldberg, "Sharecam part i: Interface, system architecture, and implementation of a collaboratively controlled robotic webcam*", IEEE/RSJ International Conference on Robots and Systems*, 2003.

[32] A. Khamis, F. J. Rodríguez & M. A. Salichs, "Remote Interaction with Mobile Robots", *Journal of Autonomous Robots* **15**, 267–281, 2003, Kluwer Academic Publishers.

[33] Archive of Online and Internet Robots. 2003. Available at: http://ford.ieor.berkeley.edu/ir/

[34] IECAT (Innovative Educational Concepts for Autonomous and Teleoperated System), http://roboticslab.uc3m.es/

[35] J. Gallagher and S. Perretta, "WWW autonomous robotics: Enabling wide area access to a computer engineering practicum**", Proceedings of the 33rd Technical Symposium on Computer Science Education**, 2002.

[36] Java Khepera Simulator, http://diwww.epfl.ch/lami/team/michel/khepsim/

[37] F. Candelas, S. Puente, F. Torres, V. Segarra, and J. Navarrete, "Flexible System for Simulating and Tele-operating Robots through the Internet", *Journal of Robotic Systems*, vol. 22, pp. 57–166, 2005.

[38] A. Khamis, M. Pérez Vernet, K. Schilling, "A Remote Experiment on Motor Control of Mobile Robots", *The 10th Mediterranean Conference on Control and Automation, MED* 2002, Lisbon, Portugal, July 9-12, 2002.

[39] I. Elhajj, A. Goradia, N. Xi, C. Kit, Y. Liu, and T. Fukuda, "Design and analysis of internet-based tele-coordinated multi-robot systems," *Journal of Autonomous Robots*, vol. 15, pp. 237 – 254, 2003.

[40] W. Xiao-Gang, M. Moallem, and R. Patel, "An internet-based distributed multipletelerobot system", *IEEE Transactions on Systems, Man and cybernetics, Part A*, vol. 33, pp. 627–634, 2003.

[41] A. Khamis, A. Abdel-Rahman, M. Kamel, "A Distributed Architecture for Mobile Multirobot Remote Teleoperation", *IEEE 2006 Workshop on Distributed Intelligent* Systems, Prague, Czech Republic, June 15 - 16, 2006.

[42] ITR, Internet Traffic Report, http://www.internettrafficreport.com/

[43] IWR, Internet Weather Report, http://www.mids.org/weather

[44] IPMA, Internet Performance and Measurement Analysis Project, http://nic.merit.edu/ipma/

[45] IPPM, Internet Protocol Performance Metrics, http://www.advanced.org/ippm.html

[46] K. Brady & T. Tarn, "Internet-based Remote Teleoperation", *In Proceedings of the 1998 IEEE International Conference on Robotics & Automation, ICRA98*, pp.65-70, Leuven, Belgium, 1998.

[47] K. Han, S. Kim, Y. Kin & K. Kim, "Internet Control Architecture for Internet-based Personal Robot", *Journal of Autonomous Robots*, **10**:135-147, 2001.

[48] W. Fung, N. Xi, W. Lo & Y. Liu, "Improving Efficiency of Internet-based Teleoperation using Network QoS", *In Proceedings of the 2002 IEEE International Conference on Robotics & Automation, ICRA02*, pp. 2707-2712, Washington, DC, 2002.

[49] D. Gillet & C. Salzman, "Real-time Interaction over the Internet", *In Proceedings, the 15th IFAC World Congress on Automatic Control (b'02)*, Barcelona, Spain, 2002.

[50] Whatis Web Page, http://whatis.techtarget.com/

[51] S. Hireche. *Control of Teleoperation Systems in QoS Communication Networks*. Thesis, Technische Universitat Berlin, 2002.

[52] N. Ando, J. Lee & H. Hashimoto, "A Study on Influence of Time Delay in Teleoperation", *In Proceeding of the 1999 IEEE/ASME Internacional Conference on Advanced Intelligent Mechatronics*, Atlanta, USA, pp. 317-322, 1999.

[53] E. Rogers and R. Murphy. *Human-Robot Interaction*. Final Report for DARPA/NSF Study on Human-Robot Interaction, 2001. Available at: http://www.csc.calpoly.edu/~erogers/HRI/HRI-report-final.html.

[54] ttcp Web Page, http://ftp.arl.mil/~mike/ttcp.html

[55] pingER Web Page, http://www-iepm.slac.stanford.edu/pinger/

[56] netperf Web Page, http://www.netperf.org/netperf/NetperfPage.html

[57] J. Kurose & K. Ross. *Computer Networking*. Addison-Wesley, 2001.

[58] D. McGovern. *Experiences and Results in Teleoperation of Land Vehicles*. Report SAND87-0646, Sandia National Lab, Albuquerque, PA, 1999.

[59] D. Bapna, E. Rollins, A. Foessel & R. Whittaker, "Antenna Pointing for High Bandwidth Communications from Mobile Robots", *In Proceedings of the 1998 IEEE International Conference on Robotics & Automation*, ICRA98, pp.65-70, Leuven, Belgium, vol.4, pp.3468-3473, 1998.

[60] C. Sayers. *Remote Control Robotics*. 1st Edition, Springer Verlag, 1999.

[61] L. Cottrell, W. Matthews & C. Logg. Tutorial on Internet Monitoring & PingER at SLAC. Available in: http://www.slac.stanford.edu/comp/net/wan-mon/tutorial.html#jitter, 2003

[62] T. Sheridan. *Telerobotics, Automation, and Human Supervisory Control*. MIT Press, Cambridge, USA, 1992.

[63] R. Alami, R. Chatila, S. Fleury, M. Ghallab & F. Ingrand, "An Architecture for Autonomy", *The International Journal of Robotics Research*, **17** (4): 315-337, 1998.

[64] R. Barber & M. A. Salichs, "A New Human-based Architecture for Intelligent Autonomous Robots", *The Fourth IFAC Symposium on Intelligent Autonomous Vehicles*, pp85-90. Sapporo, Japan, 2001.

[65] M. Boada. *Control System for Autonomous Mobile Robots based on Reactive Skills*. PhD thesis (In Spanish), Universidad Carlos III de Madrid, 2002.

[66] M. Boada, R. Barber & M. A. Salichs, "Visual Approach Skill for a Mobile Robot Using Learning and Fusion of Simple Skills", *Robotics and Autonomous Systems*, **38**:157-170, 2002.

[67] H. Rohbert, P. Sommerlad & M. Stal. *Pattern Oriented Software Architecture: A System of Patterns*. John Wiley & Sons, 1996.

[68] A. Houlb. Building User Interfaces for Object-oriented Systems. JavaWorld, July 1999, Available in: http://www.javaworld.com/javaworld/jw-07-1999/jw-07-toolbox.html

[69] D. Brugali & M. Fayad, "Distributed Computing in Robotics and Automation", *IEEE Transactions on Robotics and Automation*, **18**(4):409-420, 2002.

[70] M. Juric, I. Rozman & M. Hericko, "Performance Comparison of CORBA and RMI", *Information and Software Technology*, **42**:915-933, 2000.

[71] Q. Mahmoud. *Wireless Java*. O'Reilly, January 2002.

[72] M. Schaaf & F. Maurer, "Integrating Java and CORBA: A Programmer's Perspective", *IEEE Internet Computing*, **5**:72-78, 2001.

[73] D. M. Rivero, A. Khamis, F. J. Rodríguez & M. A. Salichs, "A Patterns-Oriented Framework for the Development of Automatic and Deliberative skills for mobile robots", The 11th International Conference on Advanced Robotics, ICAR03, University of Coimbra, Portugal, 2003.

[74] A. Khamis, F. J. Rodríguez, R. Barber & M. A. Salichs, "An Approach for Building Innovative Educational Environments for Mobile Robotics", Special Issue on Robotics Education, *International Journal of Engineering Education*, TEMPUS Publications, 2006.

[75] J. Borenstein & L. FENA, "Measurement and Correction of Systematic Odometry Errors in Mobile Robots", *IEEE Transactions on Robotics and Automation*, **12**(5), October 1996.

In: Frontiers in Robotics Research
Editor: Max A. Denket, pp. 165-199
ISBN 1-60021-097-X

Chapter 5

Adaptive Robot Controller Using Soft Computing

Dilip Kumar Pratihar*
Department of Mechanical Engineering,
Indian Institute of Technology, Kharagpur- 721 302, India

Abstract

The increasing demand for robots in various fields of application has created a thirst, to make them intelligent, interactive, cooperative and emotional as well. To reach this milestone, a lot of research is yet to be carried out, in different fields of robotics research. Motion planning and coordination of multiple robots may be considered as one of such important fields. This chapter concentrates on this field of robotics research, discusses different possible schemes using both conventional algorithmic as well as emerging soft computing approaches. It has been pointed out that the fuzzy logic and/or neural network-based controller evolved using an artificial evolution tool like genetic algorithm, could be the candidate adaptive robot controller.

Keywords: Mobile robot, motion planning, coordination, adaptive controller, soft computing

1 Introduction

Current trend in robotic research aims to build an autonomous and intelligent robot, to meet its increasing demand in manufacturing, space science, under-water applications, medical science, and others. Thus, building an autonomous and intelligent robot has become a thrust area in robotic research. Based on the collected information with the help of camera(s) and/or sensor(s), an autonomous robot should be able to take the decisions on-line, as the situation demands during its navigation. It is a complex task indeed, and to perform it successfully, the robot should have an adaptive controller, for the development of which, the principle of artificial intelligence (AI) is to be merged with that of robotics.

*E-mail address: dkpra@mech.iitkgp.ernet.in

In the year 1993, Alan Mackworth mentioned in one of his papers entitled "On Seeing Robots" that soccer could be a good test bed of robotics and AI research [1]. Thus, RoboCup (i.e., World Cup Robot Soccer) has been considered as a benchmark for studying AI techniques in robotics. The ultimate goal of the RoboCup has been set as follows [2]: "By the mid-21st century, a team of autonomous humanoid robots shall beat the human world cup champion team under the official regulations of FIFA."

To reach the target, robots should have suitable adaptive controllers, so that they can solve the following problems on-line, in a changing/ dynamic environment:

- collecting information of the environment using sensors and/or cameras,
- solving the issues related to dynamic stability,
- motion planning to determine safe and collision-free paths during navigation,
- coordination of multiple robots working in the same work-space,
- determination of control action to be performed and its realization through the motors.

Thus, it is a complex task having many open research issues. In this chapter, an emphasis will be given to the problems related to motion planning and coordination of multiple robots sharing a common work-space.

Multiple robots working in the same work-space form a multi-agent system (MAS) of robotics [3]. Robot soccer could a good domain for studying an MAS. Multi-agent systems of robotics are classified into two groups, namely centralized MAS and de-centralized MAS. In a centralized MAS, there is a supervisory computer, which controls other robots. On the other hand, in a de-centralized MAS, there is no supervisory computer and each robot performs the tasks based on its own goal, which is a function of the main goal set by the team. The present write-up will focus on de-centralized MAS. Let us take an example of soccer playing robots. Each player (i.e., robot/agent) has some goals, such as scoring goal, restricting the opponents to score goal, and others. Thus, decision of a player at a particular instant of time, depends on the positions of its both team-mates as well as opponents. Besides planning, there must be proper coordination among the team-mates, to win the game.

Many researchers have asked the question – whether planning is required in such a complex task of soccer playing. Some of them believe that the action of each player is pure reactive and there is no planning at all. Jaeger and Christaller [4] argued that in soccer playing, planning is not so much important and each player takes the decisions based on the opportunist's principle (also termed as dual dynamics). Some other investigators opine that if a robot wants to perform its tasks in optimal/near-optimal sense, planning is very much essential. But, static planning (i.e., off-line planning) may not be suitable for a dynamic environment like soccer playing. In such cases, some sort of re-planning might also be necessary. Many investigators have also given the opinion that planning reinforced by a little bit of learning (based on the previously collected information) will be the right approach, to tackle this problem more efficiently.

Hard computing [5] works based on pure mathematics and it may fail to provide online solutions to such a complex task, due to its inherent computational complexity and lack of adaptability. Soft computing [6] is a family, consisting of the tools, such as fuzzy logic, neural network, genetic algorithm and others, and their different combinations, in which precision is traded for tractability, robustness and low cost solutions. Fuzzy logic controller is a potential tool for dealing with imprecision and uncertainty. Neural network is a powerful tool for learning and adaptation, and genetic algorithm is an important tool for global optimization. Thus, soft computing could be an effective tool to develop an adaptive robot controller, which will be able to take the decision on-line, as the situation demands in a varying environment.

This chapter deals only with the issues, such as motion planning and coordination of multiple robots, out of various issues related to design and development of adaptive robot controllers. These issues are discussed below.

The rest of the text is organized as follows: Section 2 describes different schemes of robot motion planning. Some of the work on multi-robot planning and coordination are discussed in Section 3. Section 4 explains some of the work done by the author, to design optimal robot controllers using the genetic-fuzzy system. Some of the future directions of this research are discussed in Section 5. This chapter has been summarized in Section 6.

2 Robot Motion Planning

An autonomous robot should be able to plan its motion, as the situation demands in a dynamic environment. Thus, it should have a suitable motion planner. Robot motion planning could be either gross or fine in nature. Gross motion planning deals with the issues related to movement of a robot through its free-space (i.e., free from obstacles and others), whereas fine motion planning handles the compliant motion of the robot. This chapter concentrates on gross motion planning schemes only. Motion planning includes both navigation as well as manipulation. Navigation is defined as "the science of getting ships, aircraft and spacecraft from place to place, especially the method of determining position, course and distance travelled" [7]. The term – navigation has been used here, in connection with motion planning of a mobile robot. The role of a navigation scheme is to decide the route to be followed to perform a task in the optimal sense. In general, the term – manipulation is used to indicate the motion planning of a manipulator (i.e., robot with a fixed base). Thus, in a navigation problem, the search space of a mobile robot is its plane of motion, whereas in a manipulation problem, the search space is nothing but the work space of a manipulator. In this chapter, an attention will be focused on navigation problems.

2.1 Motion Planning with Complete Information

Let us take an example, in which a robot will have to make its planning in a completely known environment (also known as structured environment) containing some static obstacles. In such a situation, it can make the planning off-line, as the required information is known beforehand. It is called a *global approach*, which is a rigid one with very little information processing. It is also known as *Act-After-Thinking* process.

2.2 Motion Planning with Incomplete Information

Let us consider another situation, in which a robot will have to decide its collision-free path while navigating in the presence of some moving obstacles (i.e., the environment is un-structured). In this case, information of the environment is not known beforehand and planning is to be done on-line, based on the information collected by the sensors and/or cameras. It is a *local approach*, which is nothing but an *Act-While-Thinking* process.

2.3 Different Approaches of Robot Motion Planning

Algorithmic methods, which include both graph-based as well as analytical approaches have been developed by various investigators, to solve motion planning problems of both mobile robots as well as manipulators. These are also known as traditional methods of robot motion planning. Latombe [8] provides an extensive survey on these methods.

2.3.1 Algorithmic Methods of Robot Motion Planning

Several algorithmic methods have been proposed by various researchers. These methods include both graph-based as well as analytical approaches. Graph-based techniques, such as road map, cell decomposition, and others are developed by many investigators to determine collision-free path of a robot, while navigating in the presence of some fixed obstacles. Road map includes visibility graph [9], Voronoi diagram [10], freeway net [11], and others. *Visibility graph* is a non-directed graph, whose nodes are the initial point, final point and those vertices of the obstacles, which are visible from one another. Thus, a set of feasible paths may be obtained along which the robot can navigate after ensuring the collision-free movement. In *Voronoi diagram*, the candidate path for navigation is determined by considering the locus of points those are equidistant from at least two of the boundaries. *Freeway net* consists of a number of free ways extracted from the work-space (such as straight linear generalized cylinders) and these are connected in the form of a graph. In *cell decomposition* [12], the robot's free space is divided into a number of cells and a connectivity graph between the cells is constructed, which indicates the path to be followed by the robot. It is to be noted that these methods are computationally expensive, even for handling the navigation problems among static obstacles.

Moreover, Liu and Arimoto [13] introduced a *tangent graph method* along with A^* algorithm, to determine time-optimal, collision-free path of a mobile robot moving among some fixed obstacles. For handling the circular obstacles, tangents are drawn from the starting point to the nearest obstacle and from one obstacle to another, till it reaches the goal point. Thus, a candidate path of the robot obtained through this method could be a combination of some tangents and circular arcs. After finding all the candidate paths, an A^* algorithm is used to locate the optimal path in terms of traveling time.

In a dynamic environment (which contains some moving obstacles), visibility graph goes on changing with time. Thus, a concept of *accessibility graph* (in which an extra dimension, i.e., time is added to the visibility graph) was introduced by Fujimura and Samet [14], to solve the navigation problems of a mobile robot in the presence of some moving obstacles.

Analytical approaches of robot motion planning include potential field approach, path velocity decomposition, probabilistic approach, relative velocity approach, reactive control, and others. These methods are discussed, in short, below.

Khatib [15] introduced the concept of *potential field approach* in motion planning, in which the path of a robot is decided based on the artificial potential produced by the obstacles and the goal. The goal tries to attract the robot to-wards it, whereas repulsive potentials are generated by the obstacles. The robot moves under the action of combined attractive and repulsive potentials. This method has been used by various investigators to solve motion planning problems of robots navigating in the presence of both static as well as dynamic obstacles, separately. Although it is one of the most popular approaches of robot motion planning, it suffers from the following disadvantages:

- The performance of the method depends on the chosen potential function.
- Its solutions may get trapped into local minima, particularly when the attractive potential is balanced by the repulsive potential. This problem may also occur, while handling the concave obstacles.

Kant and Zucker [16] proposed a method known as *path velocity decomposition* for solving motion planning problems of a mobile robot in dynamic environment. In their approach, the problem is decomposed into two sub-problems, such as path planning problem and velocity planning problem. In the first phase, a collision-free path is planned considering the obstacles to be stationary, and in the second phase, the velocity of the robot is planned along the path to avoid collision with moving obstacles. Thus, the motion planning problem in dynamic environment is decomposed into two sub-problems, which significantly reduces the complexity of the problem. But, this method has the following drawbacks:

- The path planned by the robot may not be good always and it may be unnecessarily a lengthy one.
- The robot will have a jerky motion, as there is a sudden change of velocity.

As the camera and/or sensor data are generally associated with imprecision and uncertainty, planning a suitable path for a mobile robot while navigating in a dynamic environment, is a difficult task. In a partially known environment (also known as un-structured environment), a suitable model is to be developed to handle the uncertainty of it. A number of techniques had been used by various researchers to deal with the uncertainty of the environment. For example, a probabilistic approach, such as Poisson distribution was used by Sharma [17], Zhu [18] utilized a hidden Markov process, Nam et al. [19] developed a model using a random walk process, to tackle the issues related to movement of obstacles in a dynamic environment.

An attempt was made by Fiorini and Shiller [20], to solve robot motion planning problems in a dynamic environment by using the concept of *relative velocity* of the planning robot with respect to the obstacles. Thus, a dynamic motion planning problem is converted into several static motion planning problems. These static problems are then grouped into a single problem by using a vector transformation. The sets of velocity vectors are to be determined, so that the robot avoids collision with all the moving obstacles.

Reactive control scheme was proposed by Brooks [21], in which each robotic task is decomposed into a number of primitive motor behaviors, such as move-to-goal, avoid-obstacle, and others. In this scheme, each basic behavior is controlled at a different layer of the control scheme. These basic behaviors are coordinated by a central mechanism and the coordination mechanism is designed through a trial and error method. For a complex task, the number of layers increases and it becomes difficult to handle such a situation. It is important to mention that with the introduction of this scheme, a new field of robotics research known as *behavior-based robotics* has emerged. But, this scheme has the following drawbacks:

- As the primitive behaviors are hard-wired, it is unable to handle the situations, which the designer could not foresee during the design stage. Thus, it is not adaptive to the varying situations.

- It requires a large amount of computer memory, to store the basic behaviors, as a result of which, it might be computationally slow.

Computational complexity of robot motion planning in a dynamic environment has been studied by various researchers. Reif and Sharir [22] proved that motion planning in a 3-D environment containing moving obstacles is PSPACE-hard with given bounds on the robot's velocity and NP-hard without such bounds. Canny and Reif [23] showed that motion planning for a point robot in the plane with a bounded velocity is NP-hard, even when the moving obstacles are convex polygons moving at constant linear velocity without rotation.

2.3.2 Drawbacks of the Algorithmic Methods

The algorithmic methods suffer from the following drawbacks:

- Most of the algorithmic methods particularly the graph-based techniques are computationally expensive.

- A particular method might be suitable to solve only a particular type of problems. Thus, there is no versatile algorithm, which can be used to solve a variety of problems.

- Most of the methods do not have an optimization module and as a result of which, the planned path may not be optimal in any sense.

- Algorithmic methods may not be adaptive in nature, to accommodate the changes in a dynamic environment.

- These methods can not handle the issues related to imprecision and uncertainty of the collected data (using the sensors/cameras) of the environment. Thus, these methods might not be suitable for handling robot motion planning problems, particularly in an un-structured environment.

Thus, an efficient, versatile, computationally tractable and adaptive algorithm is yet to be developed, to solve the motion planning problems of robots.

2.3.3 Robot Motion Planning Using Soft Computing

Realizing the drawbacks of the conventional algorithmic approaches of robot motion planning, researchers started thinking whether soft computing approaches (which work based on biologically inspired techniques) can be utilized to solve this problem on-line, in a more efficient way. Many investigators had tested the performance of soft computing tools, to handle imprecision and uncertainty of the data related to a dynamic environment. Interested readers may refer to a recently published review paper on algorithmic and soft computing approaches of robot motion planning [24].

While navigating in a dynamic environment, a robot collects information of its surroundings, with the help of cameras and/or sensors and the collected data may not be precise in nature. A *fuzzy logic controller (FLC)*, which works based on Zadeh's fuzzy set theory [25], is a potential tool for dealing with imprecision and uncertainty. The performance of an FLC depends on its knowledge base, which consists of both the membership function distributions (also known as data base) as well as rule base. Appendix A explains the working principle of an FLC. The FLCs had been used by various researchers, to solve the navigation problems of mobile robots. In this connection, work of Beaufrere and Zeghloul [26], Takeuchi et al. [27], Martinez et al. [28], Pin and Watanabe [29], Maeda et al. [30] are worth mentioning. Moreover, Kim and Kong [31] proposed an approach, in which group intelligence of multiple robots sharing the common work-space was modeled using the FLC. Marapane et al. [32] developed an integrated autonomous multi-robots system utilizing the FLC. Realizing the strength of an FLC, many researchers utilized it, to solve the complex problem like navigation of robots, but designing the proper knowledge base of an FLC is not an easy task. Moreover, the complexity of the FLC increases with the number of variables and that of grades/terms for each variable. Thus, for solving a problem involving many variables, one may switch over to a neural network (NN).

An *artificial neural network (ANN)* was proposed by McCulloch and Pitts [33], which works based on the principle of our nervous system. The working principle of an ANN has been explained in Appendix B. Neural networks had been used by various investigators to solve motion planning problems of robots. Some of these work are explained here. Lee and Bein [34] developed an NN-based controller to determine collision-free paths of multiple cooperating robots. Botelho et al. [35] utilized a Boolean neural network for controlling a mobile robot named Khepera. The main drawback of NN lies in the fact that the training is computationally expensive and there is a chance of the solutions of a back-propagation NN, to get stuck at local minima, as it tries to minimize the error following the principle of steepest descent method.

Genetic algorithm (GA) is a population-based search and optimization technique, which works based on the mechanics of natural genetics and Darwin's principle of natural selection. The concept of a GA was introduced by Prof. John Holland of the University of Michigan, USA [36] (although this book was published in 1975, the idea was proposed in the year 1965). Appendix C explains the working principle of a GA. The GA had been used by various researchers as an optimization tool, to solve the motion planning problems of robots. Mitchell [37] utilized a GA along with a grid search and visibility graph search methods to determine collision-free paths for the robot. Zhao et al. [38] solved the path planning problem of a mobile manipulator system by using a GA.

A GA had also been utilized to improve the performance of a fuzzy logic controller and this combined scheme is known as the *genetic-fuzzy system*. Pratihar [39] had developed a genetic-fuzzy system, to plan both the path as well as gait of a six-legged robot, in an optimal sense. Moreover, Pratihar and Bibel [40] was successful to determine time-optimal and collision-free paths for multiple robots working in the same work-space by using the genetic-fuzzy system.

Combined *NN-fuzzy systems* had also been developed by various researchers, to eliminate their individual limitations and get the advantages of both the tools. Some of the applications of the NN-fuzzy systems are mentioned here. Li et al. [41] developed a neuro-fuzzy system for controlling a mobile robot in an unknown environment. The environment was modeled by using a neural network. Each task was decomposed into a number of basic behaviors and the coordination problem of these behaviors was solved by using a fuzzy reasoning. Ng and Trivedi [42] designed a neuro-fuzzy controller for solving mobile robot navigation and multi-robot convoying problems. They developed a fuzzy logic-based controller, which was fine-tuned on-line, by adjusting the weights of the neural network. Topalov et al. [43] developed an approach for navigation of a wheeled mobile robot in a partially-known environment. The NN determined a collision-free path of the robot and a fuzzy logic-based strategy was used, when the robot faced an unexpected obstacle. Acosta et al. [44] proposed a neuro-fuzzy system to plan collision-free path of a mobile robot. In their approach, the knowledge base of the fuzzy logic controller was tuned by using an NN.

Some investigators had also tried to solve this complex problem of motion planning by using the concept of biological adaptation. Learning and evolution are two forms of biological adaptation. Learning occurs in one's life-time, whereas evolution may occur through several generations. It is also understood that learning and evolution may help each other, to finally accelerate the speed of biological adaptation. Realizing this fact, investigators working in this field of robotics research, started thinking whether an artificial learning tool (i.e., neural network) can be combined with an artificial evolution tool (i.e., GA), to develop an adaptive robot controller and as a result of which, a new field of robotics research known as *evolutionary robotics* has emerged. Interested readers may refer to the review paper on evolutionary robotics [45].

It is important to mention that the robot controller designed using the principle of behavior-based robotics (discussed in Section 2.3.1), may not be adaptive in nature. Thus, there is a need for a technique, by which the robot will be able to adapt new behaviors automatically, as the situation demands. Evolutionary robotics may provide a feasible solution to this problem.

In evolutionary robotics, the basic behaviors required to solve a particular task emerge automatically through evolution, due to the interactions of the robot with its environment, and the coordination of behaviors is achieved through a self-organizing process (rather than by an explicit design). The designer designs the fitness function, which measures the ability of a robot to perform a desired task. Thus, the robot and its environment form a highly dynamical system, in which developing a suitable controller for the robot, is the prime goal of evolutionary robotics. A lot of research had been carried out by various researchers, in evolutionary robotics. Some of these work are mentioned in short, here. Floreano and Mondada [46] was successful in developing an adaptive NN-controller for the Khepera. Nolfi et al. [47] developed an adaptive NN-controller using a GA, on computer simulations and tested it on a real robot – Khepera. Moreover, some research work had been conducted at the University of Sussex, UK, on this area [48]. Evolutionary robotics could be a candidate tool, to develop an adaptive robot controller, but the main drawback of this approach lies in the fact that the GA has a slow convergence rate.

The evolution of primitive/basic behaviors through planning (reinforced by learning) could be an interesting approach, to solve this problem. It is important to mention that learning can be provided either off-line or on-line based on some training data. During off-line learning, training is provided based on the previously collected data, whereas in case of on-line learning, the training data are to be collected during the execution of tasks. Thus, in case of on-line learning, the controller may not behave in an optimal sense initially, but it will gradually improve its performance. It is also felt that a robot should adapt its control algorithm, on-line. Thus, planning supported by on-line learning using the principle of evolutionary robotics could be an appropriate approach to solve this complicated task, in which a faster GA, such as micro-GA [49], could be used as an evolution tool, to evolve NN/FL-based adaptive robot controller.

3 Multi-robot Planning and Coordination

This chapter focuses on a de-centralized multi-robot (agent) system. Let us take an example of soccer playing robots. Two teams are playing football and each team consists of eleven players. Each player (i.e., agent) tries to perform its task in the optimal sense. Thus, each player has a goal, which is the function of the main goal set by the team (i.e., to score the maximum number of goals and restrict the opponents to score goals), and responsibility of a team is distributed among the team-mates. This task assignment itself is a difficult problem. Before assigning the tasks to different agents, classification of tasks is to be done, because an agent might be able to perform a task in the optimal sense, if and only if, it is the specialist of that task. Thus, a specialist goal keeper may not perform well as a striker, and so on. After the classification is over, the tasks are to be assigned to each player.

To perform the task in an efficient way, there must be proper coordination among the team-mates. An agent should be able to coordinate/cooperate with its team-mates and at the same time, it should be able to predict the behavior of its opponents, and if possible, create some hurdles for its opponent without colliding physically. Thus, it is not as simple as a collision-free, time-optimal path planning of multiple robots, working in a common dynamic environment [40], but something more complex than that.

Several attempts were made by various researchers, to develop a suitable algorithm for solving the path planning and/or coordination problems of multiple robots sharing the common work-space. *Plan-merging algorithm* was suggested by Alami et al. [50], to solve cooperation problems of a large of autonomous robots used for load transport in a structured environment. In that scheme, each cooperating robot merges its plan incrementally into a set of already coordinated plans. Svestka and Overmars [51] developed *probabilistic road-map planners* to solve motion planning problems of multiple car-like robots working in the same work-space. Moreover, a *potential field-based randomized path planner* was designed by Barraquand et al. [52], to solve the planning problems of multiple robots. Vail and Veloso [53] presented a framework for task assignment and coordination of multiple robots in a soccer domain, using the potential field approach. Multiple robots sharing the common work-space should be able to make their plan and coordinate among themselves in an adaptive way on-line, while working in a dynamic environment. Thus, there are still many open research issues in multi-robot planning and coordination, to develop an adaptive and robust control algorithm. Soft computing approaches might be helpful to develop such algorithms.

To have proper coordination, while working in a common work-space, the robots should be able to communicate with each other. A human-being is intelligent, interactive, cooperative but at the same time, he is emotional as well. Thus, the issues related to emotion of a robot are to be dealt. Like a human-being, a robot should also be intelligent, interactive, cooperative and emotional. Recently, Prof. Fukuda of Nagoya University, Japan, has done a collaborative work with Prof. Graefe of Munich University, Germany, and they are able to develop an intelligent, interactive robot with emotion, using the principle of soft computing [54].

4 Work Done by the Author Using the Genetic-fuzzy System

The performance of a fuzzy logic controller (FLC) depends on its knowledge base (KB), which consists of both the membership function distributions (i.e., database) and rule base. It is difficult to design the proper KB of an FLC, so that it will perform in the optimal sense. Thus, tuning is required (based on a set of training data) to develop an optimal KB of an FLC by using an optimization tool. In the genetic-fuzzy system, the performance of an FLC

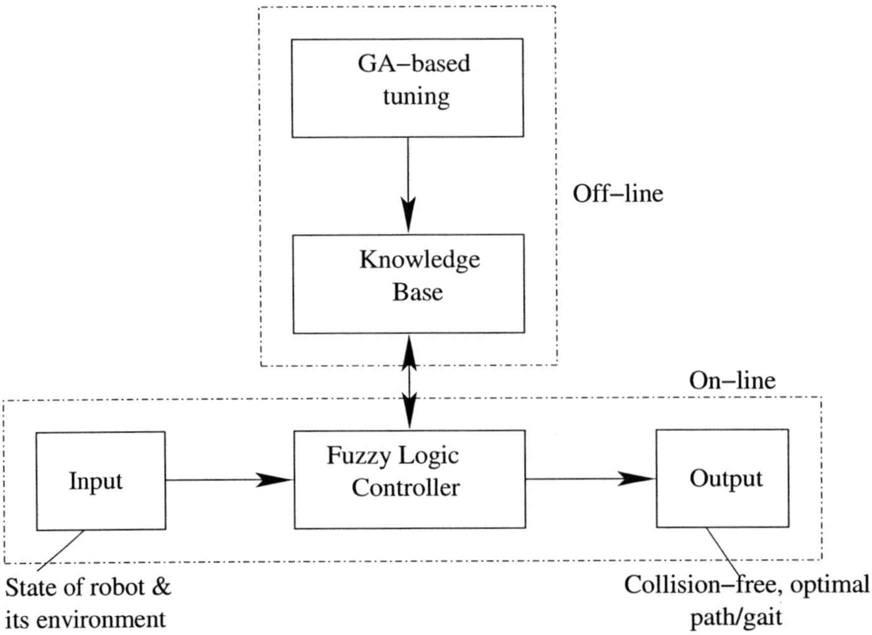

Figure 1: A schematic diagram showing the genetic-fuzzy system.

is improved by a GA-based tuning. Fig. 1 shows the schematic diagram of a genetic-fuzzy system, in which the GA-based tuning is done off-line. Once optimized, the FLC will be able to perform in an optimal/near-optimal sense. Some of the work done by the author in this field of research are discussed below.

4.1 Path Planning of a Point Mobile Robot [55]

The problem may be stated as follows: A mobile point robot has to navigate from an initial position to a final position in minimum traveling time, by avoiding the collision with a set of moving obstacles. Each moving obstacle is represented by its bounding circle. Three different approaches were developed. In the first approach, the FLC was designed manually based on the knowledge of the designer regarding the physical problem and no optimization was carried out to further improve the performance of the FLC. In Approach 2, a genetic-fuzzy system was developed, in which the manually-constructed data base and rule base of the FLC were optimized by using a GA. In the third approach (i.e., another way of designing the genetic-fuzzy system), the whole task of designing a proper KB of the FLC was given to the GA and the designer had a passive role. Fig. 2 shows the results of the three approaches.

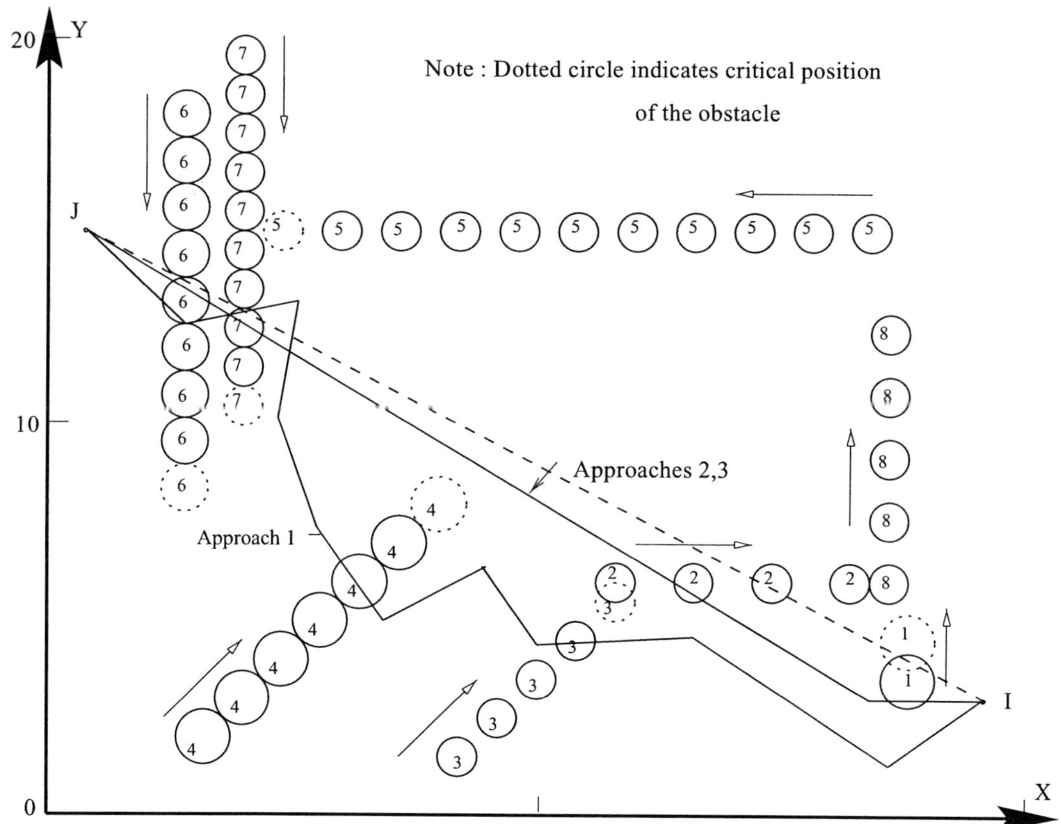

Figure 2: Optimal path planning of a point robot navigating among some moving obstacles using the genetic-fuzzy system [55].

The genetic-fuzzy system was found to perform better than the manually-designed FLC, as expected [39].

4.2 Optimal Ditch-Crossing Gait of a Six-Legged Robot [56]

The problem considered in this study was as follows: During navigation, a six-legged robot will have to cross a ditch with minimum number of legs on the ground (to avoid a deadlock situation) having the maximum average kinematic margin (to ensure a higher speed). Moreover, its stability margin should always be positive to ensure static stability. Fig. 3 shows

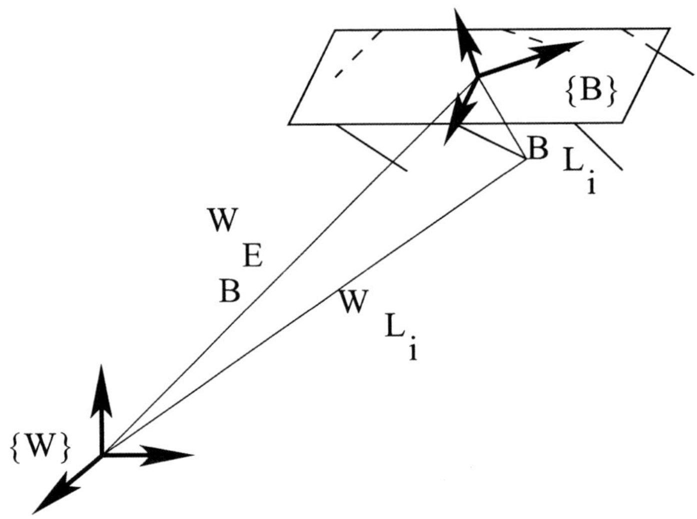

Figure 3: A schematic diagram showing the world and body frames of a six-legged robot.

the schematic diagram of a six-legged robot and the world-coordinate frame W and body-coordinate frame B assigned to it, for the purpose of analysis. Each leg of the hexapod was controlled by a separate FLC. Thus, there were six FLCs running in parallel, to control the robot. The performances of the FLCs were improved by using an off-line GA-based tuning. The optimal gaits generated by the hexapod are shown in Fig. 4. The genetic-fuzzy system was found to outperform the ordinary FLC.

4.3 Optimal Turning Gait of a Hexapod [57]

An attempt was also made by the author, to generate optimal gaits of a hexapod, which was supposed to take a circular turn through some angle in either clockwise or anti-clockwise direction, in an optimal sense (with minimum number of ground-legs having the maximum average kinematic margin) after maintaining its static stability.

Fig. 5 shows the optimal turning gaits of the six-legged robot, as developed by the genetic-fuzzy system. Once again the genetic-fuzzy system maintains its supremacy over the ordinary FLC (the gaits generated by the ordinary FLCs are not shown in this write-up).

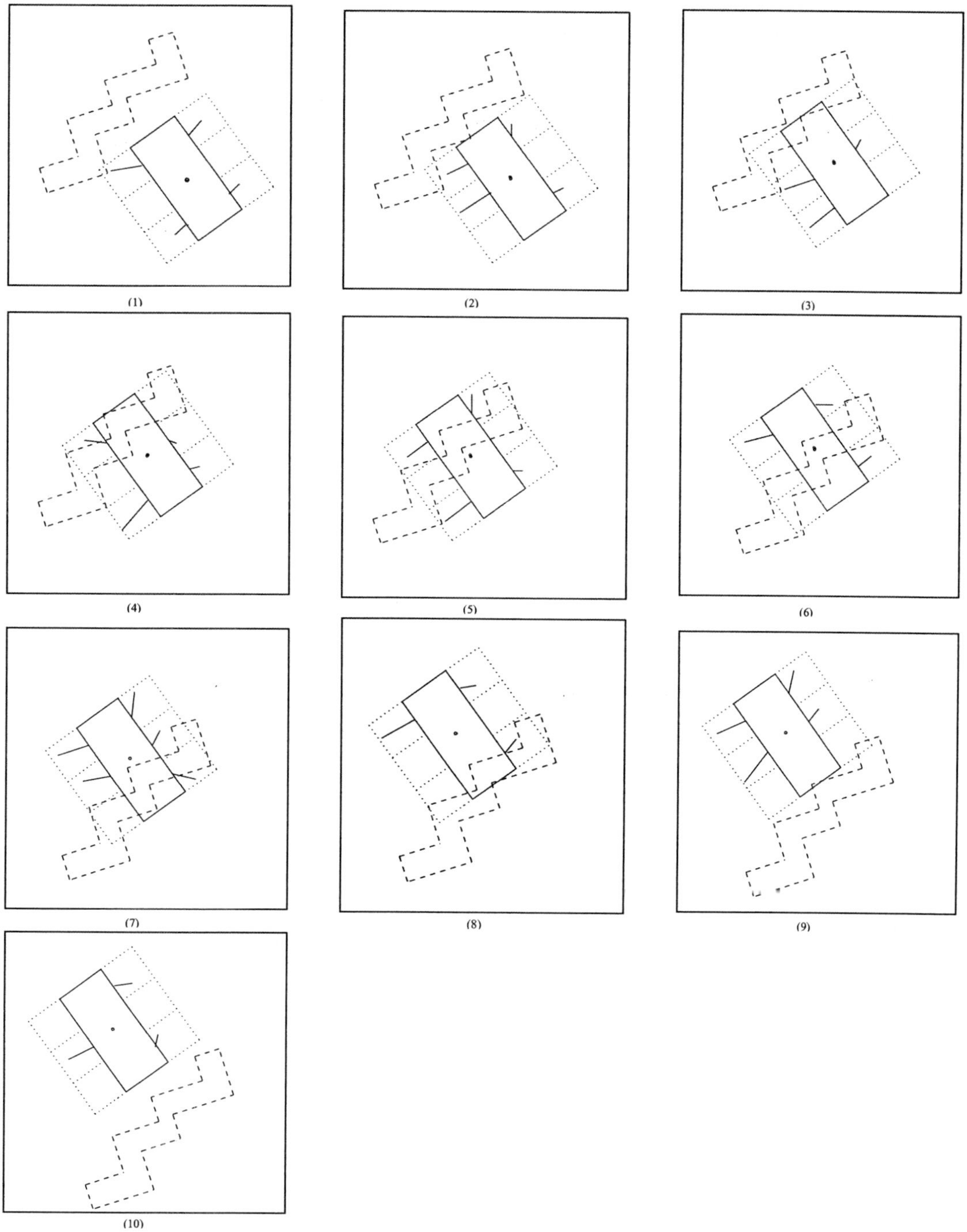

Figure 4: Optimal ditch-crossing gaits of a six-legged robot obtained using the genetic-fuzzy system [56].

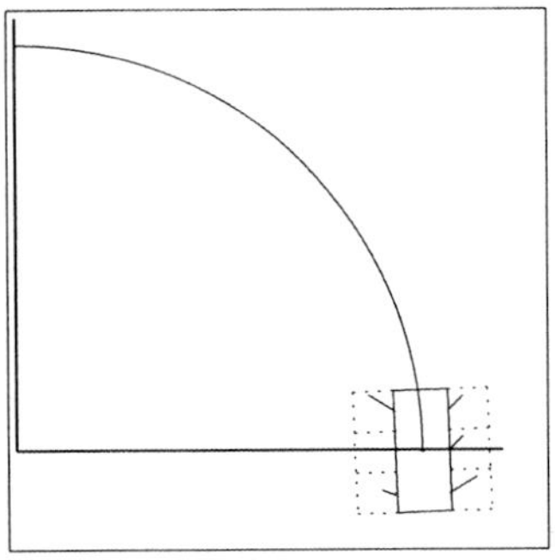
[1]

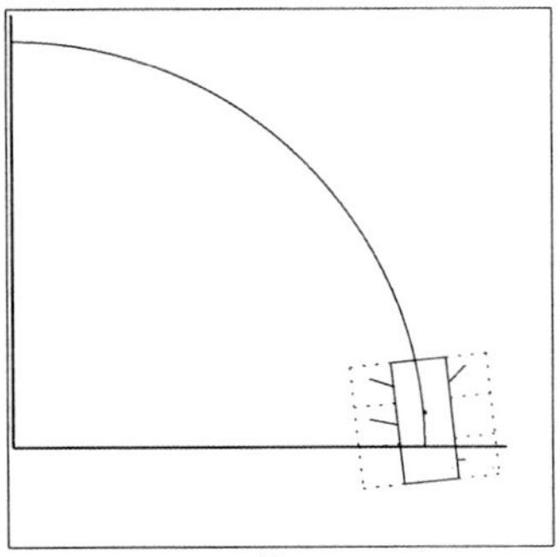
[2]

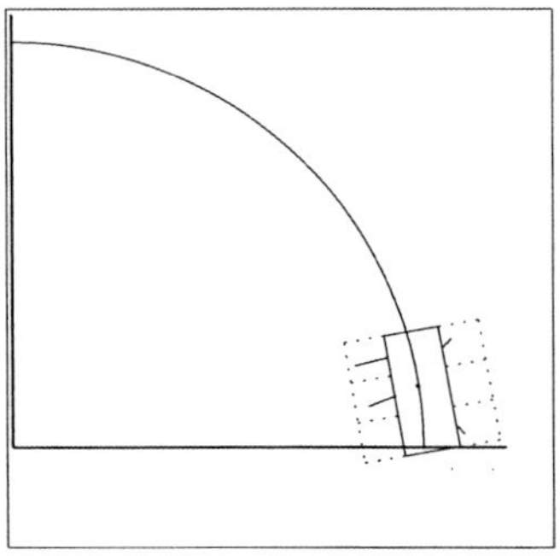
[3]

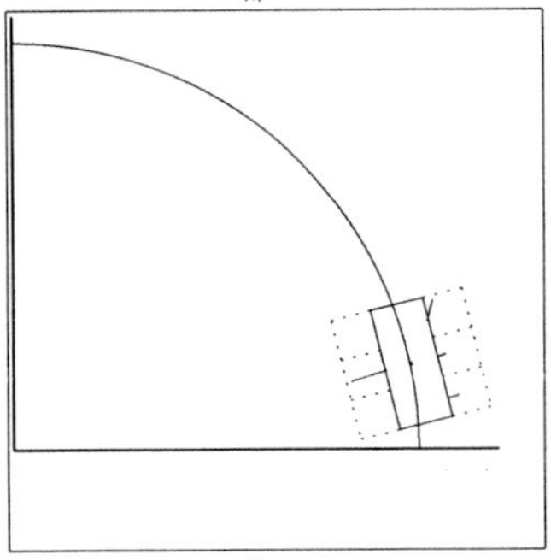
[4]

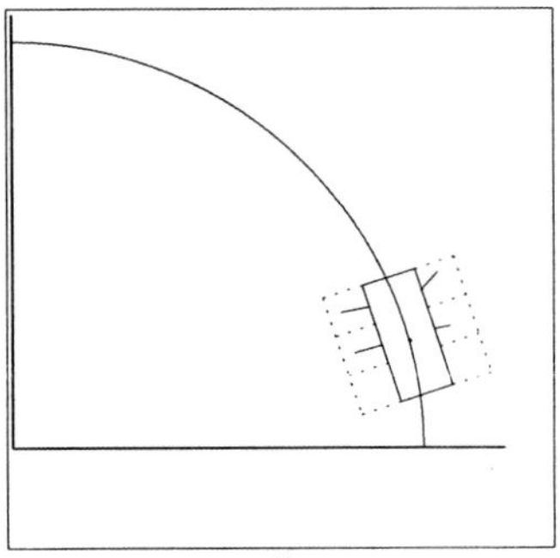
[5]

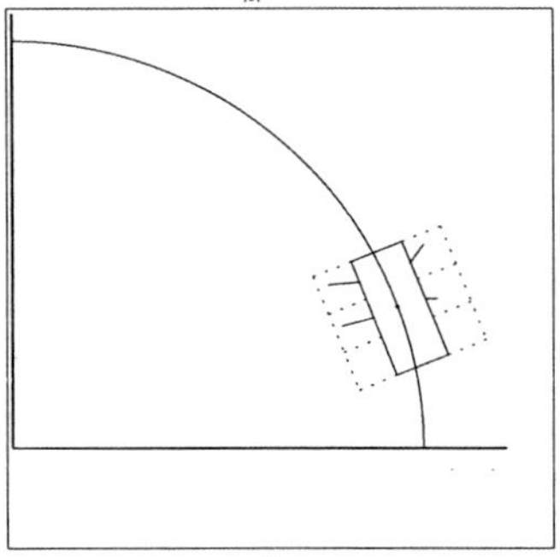
[6]

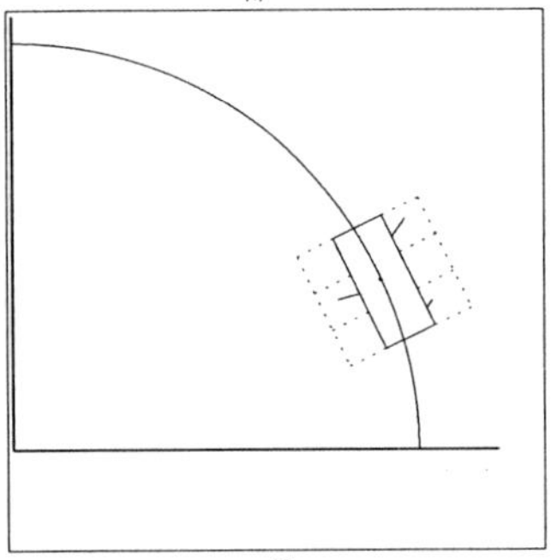
[7]

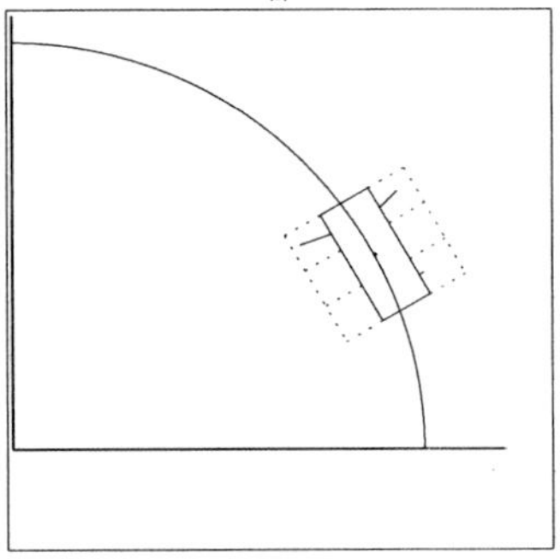
[8]

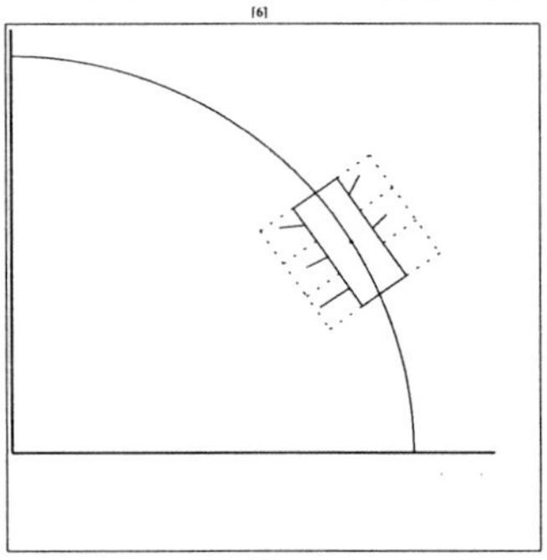
[9]

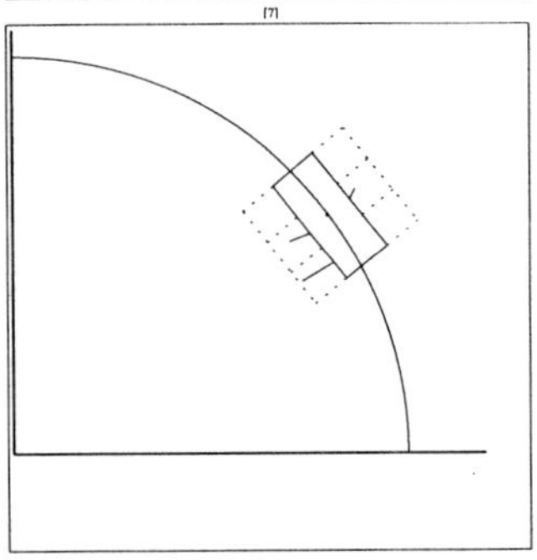
[10]

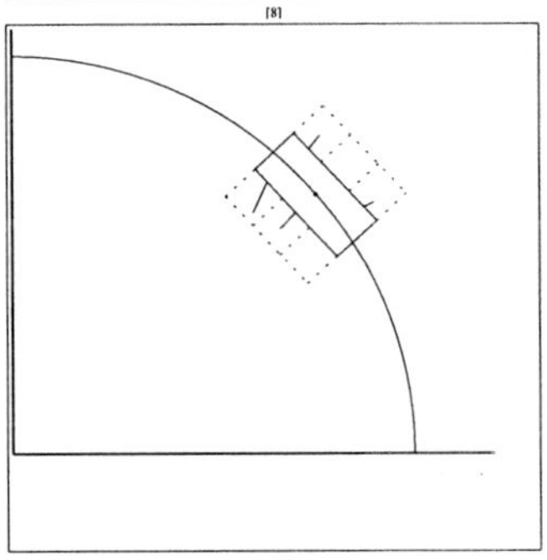
[11]

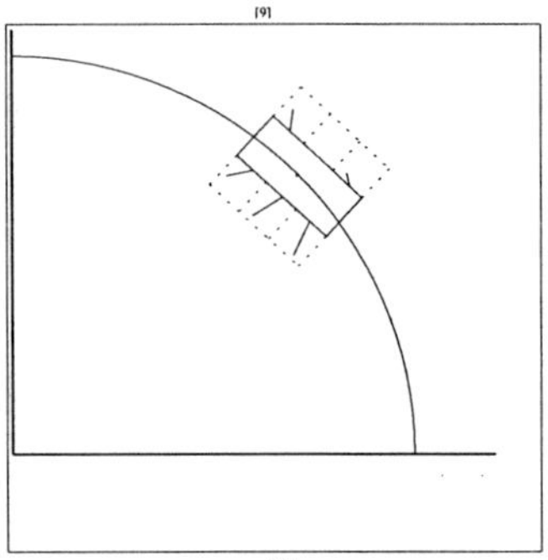
[12]

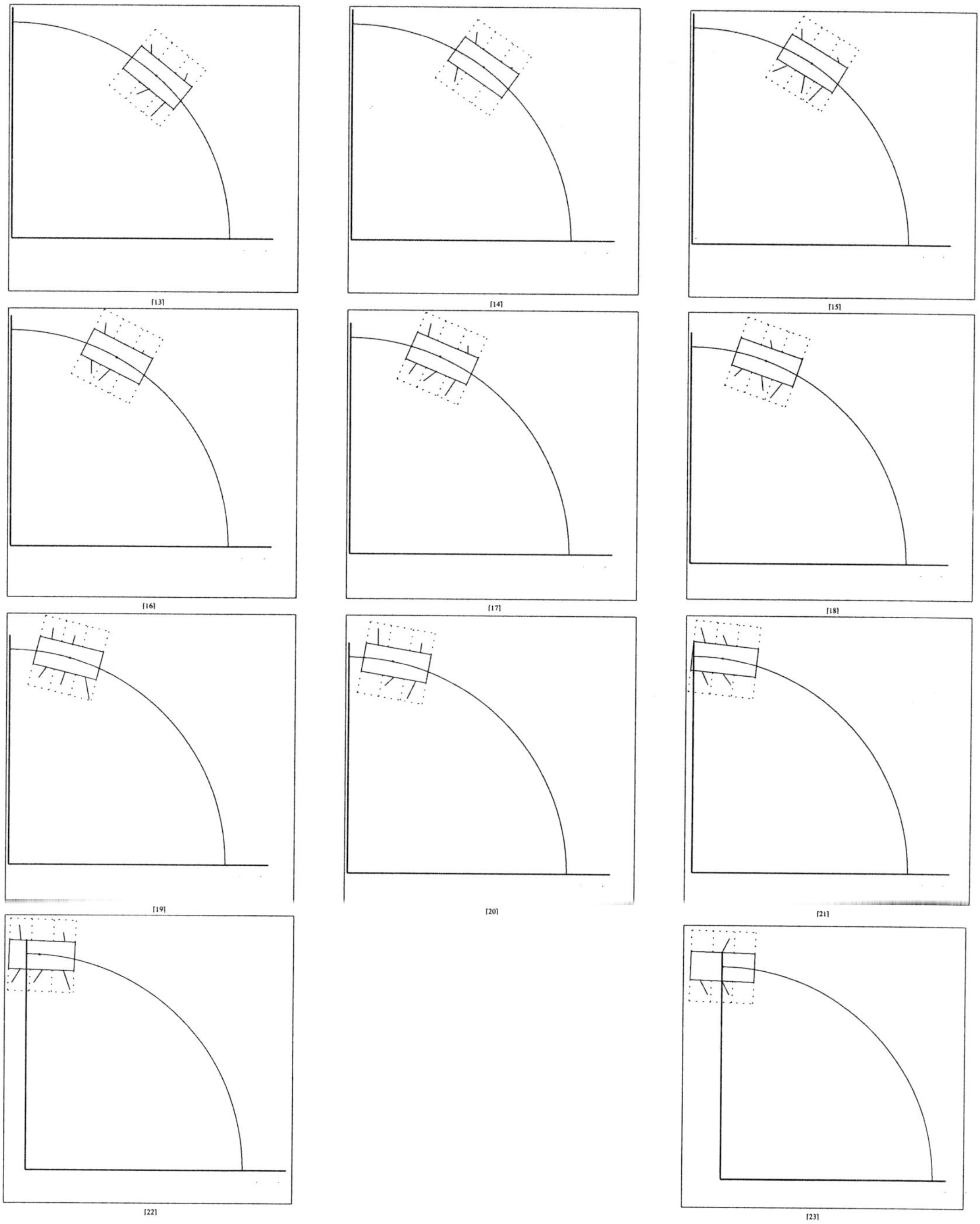

Figure 5: Optimal turning gaits of a six-legged robot obtained using the genetic-fuzzy system [57].

4.4 Combined Path and Gait Generations Simultaneously of a Six-Legged Robot [58]

In practice, a legged robot generates its path and gait simultaneously. Thus, it is a complicated problem and determining a suitable algorithm for on-line control of the robot is a difficult task.

This problem of combined path and gait generations was solved using the FLCs. Fig. 6 shows the optimal path and gait of the hexapod generated by using the genetic-fuzzy system, which was able to generate better path and gait for the hexapod compared to the ordinary FLCs. Moreover, the CPU time of the optimized FLC was found to be much less than a second, in an HP9000, K200 machine. Thus, it might be suitable for on-line implementations.

4.5 Optimal Path Planning of Multiple Robots [40]

An attempt was also made by the author, to plan optimal paths of multiple cooperating robots working in the same work-space. These problems were solved by using the ordinary FLCs and the genetic fuzzy system, separately. As expected, the genetic-fuzzy system was able to perform better than the ordinary FLCs. Fig. 7 shows the results of twelve cooperating robots, obtained by using the genetic-fuzzy system. It is important to mention that in this work, the issues related to cooperation of multiple robots have not been dealt, in detail.

4.6 Optimal Ditch Crossing Gait of a Two-Legged Robot [59]

The author also carried out some work along with one of his students, on ditch crossing gait generation of a two-legged robot. A two-legged robot will have to cross a ditch in minimum time after maintaining both its static as well as dynamic stability. The genetic-fuzzy system was able to generate better gaits compared to the ordinary FLCs. Fig. 8 shows a schematic diagram, in which a two-legged robot is crossing a ditch in the optimal sense, after maintaining its stability.

As discussed above, the genetic-fuzzy system was able to plan both the path as well as gait (of a legged robot) efficiently and almost in all the test cases (which were considered at random), it performed better than the ordinary FLCs. But, the most important thing to mention here is its adaptability. It was observed several times that the genetic-fuzzy system is able to evolve a new and efficient control action, while handling the different test cases (which were not considered during its training phase). Thus, the genetic-fuzzy system can develop an adaptive controller, which can tackle the varying situations. Moreover, the computational complexity of the developed algorithm is found to be such that it can be implemented on-line.

5 Future Directions

After achieving interesting results with the genetic-fuzzy system, the author started thinking whether the adaptive fuzzy logic-based robot controller can be developed by merging the fuzzy logic technique with other members of the soft computing family, and as a result of which, some other combined techniques, such as neuro-fuzzy, genetic-neuro-fuzzy,

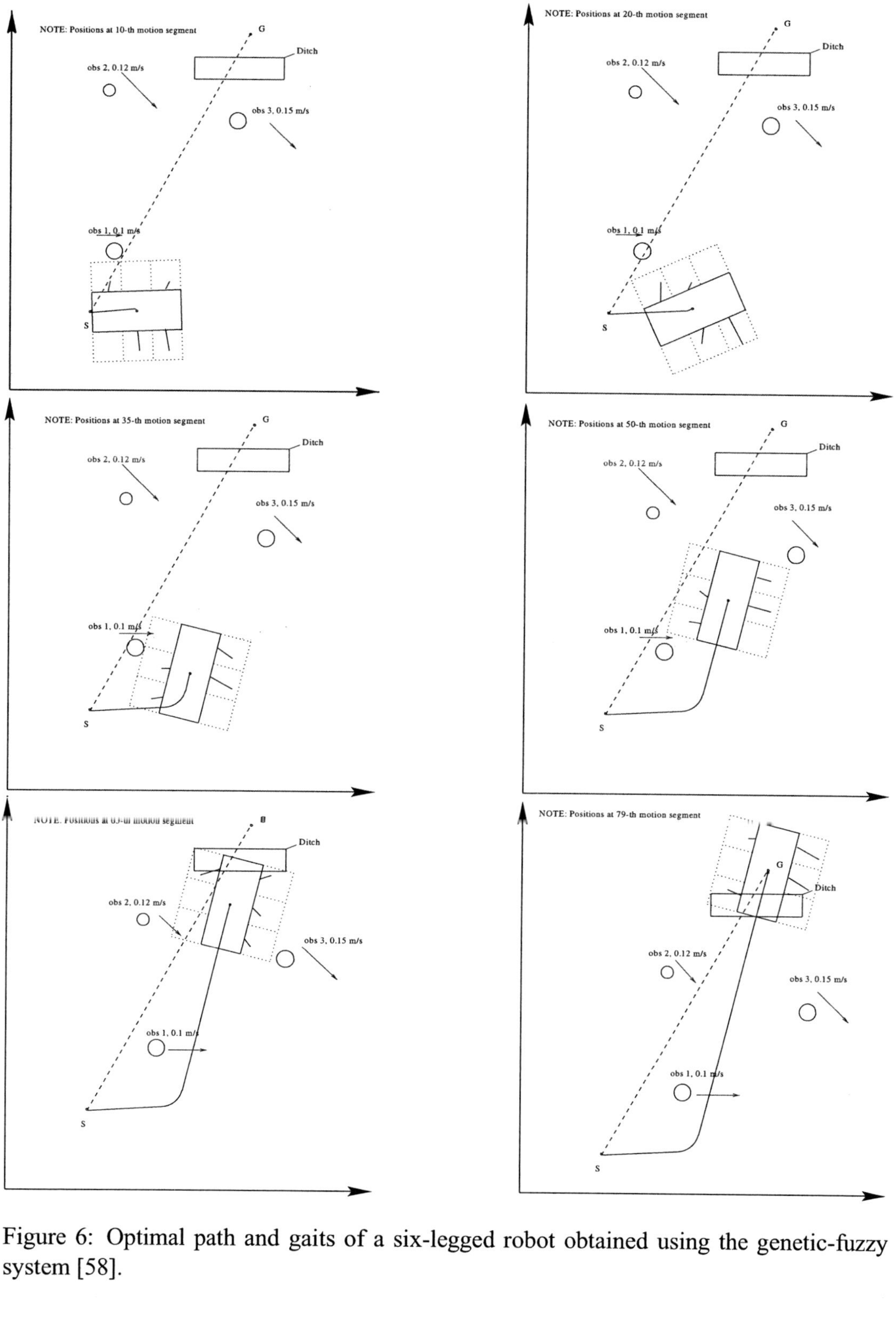

Figure 6: Optimal path and gaits of a six-legged robot obtained using the genetic-fuzzy system [58].

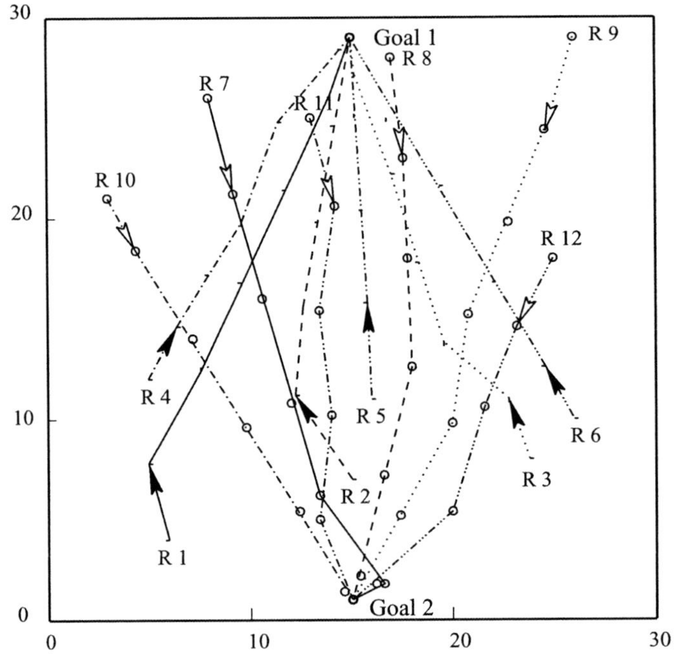

Figure 7: Optimal path planning of twelve cooperating robots using the genetic-fuzzy system [40].

and others are in process of development. Fuzzy logic controllers are becoming more and more popular, now-a-day, for their simplicity and inherent strength for handling the issues related to imprecision and uncertainty. But, its complexity increases with the number of variables, as a large number of rules are to be designed to handle the problems involving many variables. For example, a fuzzy logic controller may not be able to handle the problem related to weather forecasting properly. Thus, for a more complicated problem involving many variables, one may switch over to neural network (NN)-based controller also. A back-propagation neural network may serve the purpose, but the combined GA-NN could be a more interesting approach. Experiments are to be carried out (at least on computer simulations), to compare these candidate controllers, in terms of their performance, computational complexity and adaptability. The author is, at present, working on these issues along with one of his students. Some of these combined techniques are discussed below, in short.

5.1 Neuro-fuzzy System

In this system, the knowledge base of a fuzzy logic controller (FLC) is optimized by using a back-propagation neural network (BPNN). In this scheme, the FLC will be represented by using the structure of an NN. Fig. 9 shows the schematic diagram of the neuro-fuzzy system. The NN consists of five layers. The first layer is the input layer, which contains two neurons to represent two inputs – I_1 and I_2 (one for each input). The second layer is known as the fuzzification layer and the weights between the first and second layers (i.e., $[V]$) represent the information regarding the membership function distributions of the input

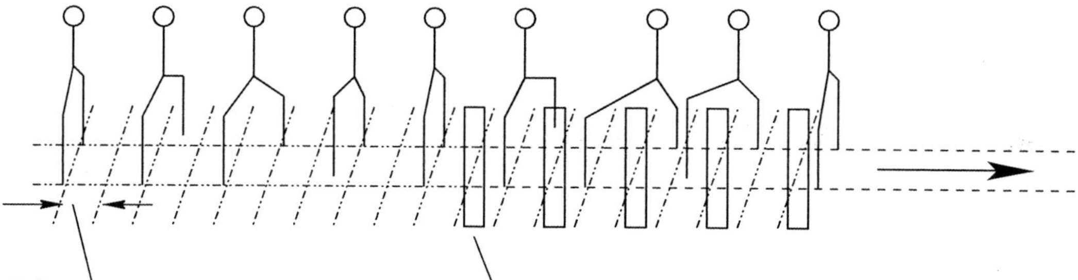

Figure 8: Optimal ditch-crossing gait of a two-legged robot.

variables. The output of the second layer is nothing but a set of membership function values. The third layer indicates the different input conditions, i.e. rules, and it also performs the logical AND operation. The logical OR operation is carried out in the fourth layer. The fifth, i.e., the output layer performs the de-fuzzification operation. The weights between the fourth and fifth layers, i.e., $[W]$ carry information of the membership function distributions of the output variables. This FLC expressed in the form of an NN can be used to solve the problems related to motion planning and coordination of multiple robots sharing a common work-space. A back-propagation algorithm may be used to optimize the knowledge base of the FLC, while solving a number of training scenarios created at random. As the back-propagation algorithm works based on the principle of steepest descent method, there is a chance of the solutions for getting trapped into local minima. Thus, the back-propagation algorithm may be replaced by a global search and optimization tool like genetic algorithm. Thus, a genetic-neuro-fuzzy system can also be developed, to solve the said problem.

5.2 Genetic-Neural System

Realizing the fact that it is difficult to develop an adaptive robot controller through explicit design, investigators working in this field of robotics research, have started started thinking whether an artificial evolution tool like genetic algorithm, can be used to evolve an adaptive controller. This field of robotics research is known as evolutionary robotics, the principle of which has been discussed earlier (refer to Section 2.3.3). The navigation and coordination problems of multiple robots may be solved by using an NN controller (refer to Appendix B). The robots collect information of the dynamic environment with the help of sensors/cameras. Based on this collected information, the inputs are determined and fed to the NN-controller. The controller decides the actions to be taken by the robots, to tackle that particular situation. During training of the NN-controller, a binary-coded GA is used to represent the weights and topology of the controller. The GA starts with a population of binary strings, created at random. The working principle of a GA has been explained in Appendix C. Fig. 10 shows the method of calculation of the fitness of the GA-strings. The designer defines the fitness function of the GA-string (in terms of traveling time) in such a way that it can measure the ability of a robot to perform a desired task. Training is provided to the NN-controller by using a number of training scenarios created at random. During training, the GA tries to find an optimal controller through search and optimization. Once

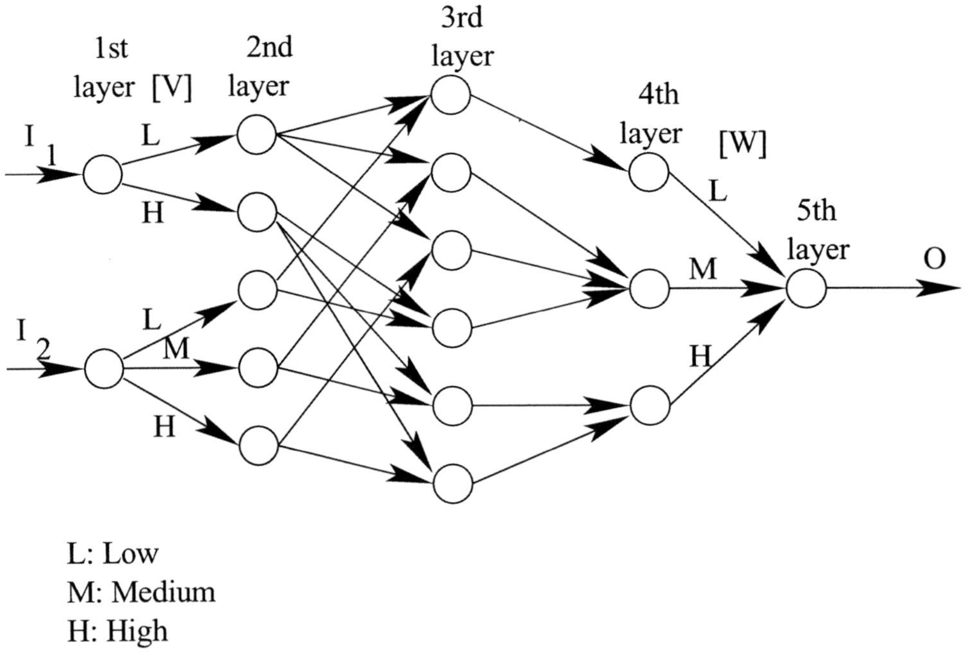

Figure 9: A schematic diagram of the neuro-fuzzy system.

trained, the NN-controller may behave in an optimal sense and show some adaptability while solving some new test scenarios.

6 Summary

To reach the milestone by mid-21st century, in which a team of autonomous humanoid robots will beat the human world cup champion team, a lot of research is to be carried out, to design and develop an adaptive controller for each robot. While working in an un-structured environment, multiple autonomous and intelligent robots working in the common workspace, should be able to perform the following tasks on-line (preferably in optimal sense): (i) gathering information of the environment, (ii) motion planning and coordination among themselves, (iii) solving their own stability problems considering dynamics, and others. Thus, it is complicated task, in which the issues related to imprecision and uncertainty are also involved. Many researchers are working, at present, on each of these sub-problems, but a lot of work is yet to be done.

This chapter concentrates only on the issues related to motion planning and coordination. Several trials were made by various researchers, to solve these problems using the conventional algorithmic approaches. They were successful partially, but only in a limited domain. Each algorithm has its inherent limitations. For example, some of these algorithms were computationally expensive, some were not versatile and adaptive in nature. Moreover, the issues related to imprecision and uncertainty could not be solved properly, although several trials were made. Thus, a versatile and computationally tractable algorithm is yet to be developed, which is adaptive in nature.

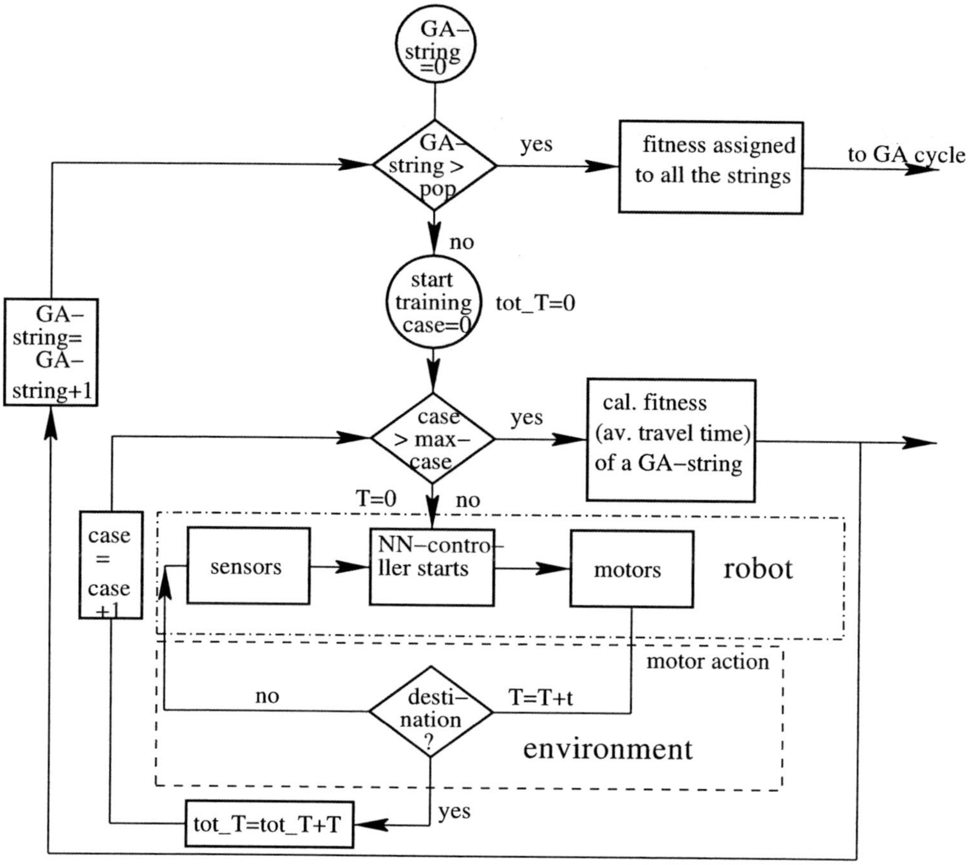

Figure 10: A schematic diagram to show the method of calculating the fitness of the GA-strings [15].

Recently, soft computing has entered into this field of robotics research. Many investigators have recently applied soft computing to solve these problems of robotics. They have tried to develop either fuzzy logic or neural network-based robot controllers using the potential of artificial evolution tools like genetic algorithm, genetic programming, and others, to inject adaptability in the control algorithm. It is important to mention that a fuzzy logic-based control algorithm might be suitable to handle a problem involving only a few variables, but for tackling a problem having many variables, one may switch over to neural network-based controllers. Thus, before selecting a suitable control algorithm (either FL-based or NN-based) to solve a particular problem, comparisons are to be made of these control algorithms (such as *genetic-fuzzy system, neural-fuzzy system, genetic-neural-fuzzy system, genetic-neural system*, and others), in terms of their efficiency, ability in developing adaptive controller, computational complexity, and others. It might be advisable to make this comparison on computer simulations, before going for carrying out experiments using real robots.

In behavior-based robotics, a task is divided into a number of basic behaviors and these basic behaviors are coordinated using a central mechanism. This approach is rigid in the sense that if the robot faces a situation which is not considered during its design stage, it may fail. Thus, it does not have any adaptability. Soft computing may be used to bring adaptability in this approach, so that new behaviors and their coordination mechanism will be evolved automatically in an un-structured environment.

Let us take the example of soccer playing robots. Each player (i.e., intelligent robot or agent) of a team will have to perform a number of tasks. There might be some overlapping of the tasks also, to be performed by two robots of a team. Moreover, a player should be able to predict the movement (behaviors) of the opponent players. Thus, the tasks to be performed by a player, may be divided into a large number of basic behaviors and a suitable mechanism is to be evolved to coordinate these behaviors. It is a difficult task and an autonomous robot might be able to do the planning accordingly, if it would have proper learning. Learning can be provided either off-line or on-line based on some training data. It is also felt that on-line learning could be more effective (to bring adaptiveness in the control algorithm) compared to off-line learning. Thus, planning reinforced by on-line learning could be the right approach to solve this problem. Soft computing tools can be utilized during these learning and planning phases. Either FL or NN-based planner can be developed to tackle these problems and an off-line/on-line learning using an evolutionary computing tool or a genetic algorithm may further improve the performance of the planner. It is to be noted that the genetic or evolutionary algorithms are generally slow and they can be used for off-line learning, but for carrying out on-line learning, a faster GA, namely micro-GA might be suitable.

Besides intelligence, emotion plays an important rôle in the decision making process of a human-being. The issues related to modeling of human emotion artificially, will be searched in future, using the principle of soft computing. Thus, in future, soft-robots (i.e., robots having adaptive controllers developed based on soft computing) will be developed, which are intelligent, autonomous and emotional too. But, many miles are yet to be run, before we reach.

Acknowledgment

The author was lucky to get advice from Prof. Amitabha Ghosh & Prof. Kalyanmoy Deb of IIT Knapur, India, and Prof. Wolfgang Bibel of Dramstadt University of Technology, Germany, while carrying out research work on this field of robotics. He gratefully acknowledges the help and advice of them. Moreover, he thanks some of his students, namely Balvinder Singh, Nirmal Baran Hui, for their fruitful discussions on some of the topics.

Appendix A

A Fuzzy Logic Controller

The concept of fuzzy set theory was first introduced by Prof. Zadeh [25], in the year 1965. A fuzzy logic controller (FLC) [60], one of the most successful applications of fuzzy set theory, is a powerful tool for handling imprecision and uncertainty. An FLC mimics the behavior of a human operator, who would be able to control a process. It does not depend on the extensive mathematical formulation of the problem. Thus, it can be used to control a complicated process, which is difficult to model mathematically or the mathematical formulation is highly non-linear. In FLC, the variables are expressed in terms of membership function distributions and human knowledge is expressed in terms of linguistic control rules. The performance of an FLC depends on both the membership function distributions of the variables (also known as data base) as well as rule base. Different types of membership function, such as triangular, trapezoidal, Gaussian, bell-shaped, sigmoidal, and others, are generally used to represent the variables.

An FLC consists of four modules, namely a fuzzy rule base, a fuzzy inference engine, fuzzification and defuzzification. The following steps are involved in the working cycle of

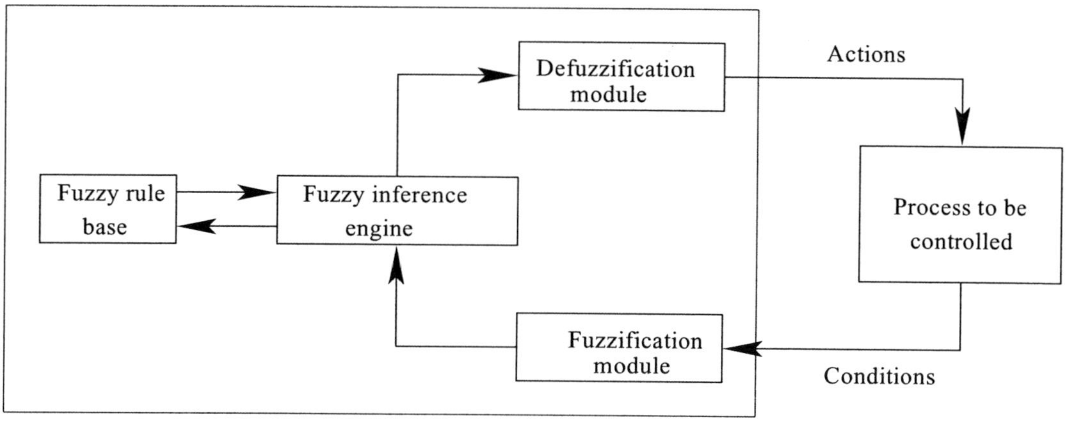

Figure 11: A schematic diagram showing the working cycle of an FLC.

an FLC (refer to Fig. 11):

- The variables (both condition as well as action) needed to control a particular process are chosen and measurements are taken of all the condition variables.
- The measurements taken in the previous step are converted into appropriate fuzzy sets to express measurement uncertainties. This is known as *fuzzification*.
- The fuzzified values are then used by the inference engine to evaluate the control rules stored in the fuzzy rule base and a fuzzified output is determined.

- The fuzzified output is then converted into a crisp value. This conversion is called a *defuzzification*. The defuzzified values represent the actions to be taken by the FLC in controlling the process.

The fuzzy reasoning process is illustrated in Figure 12. For simplicity, we assume that

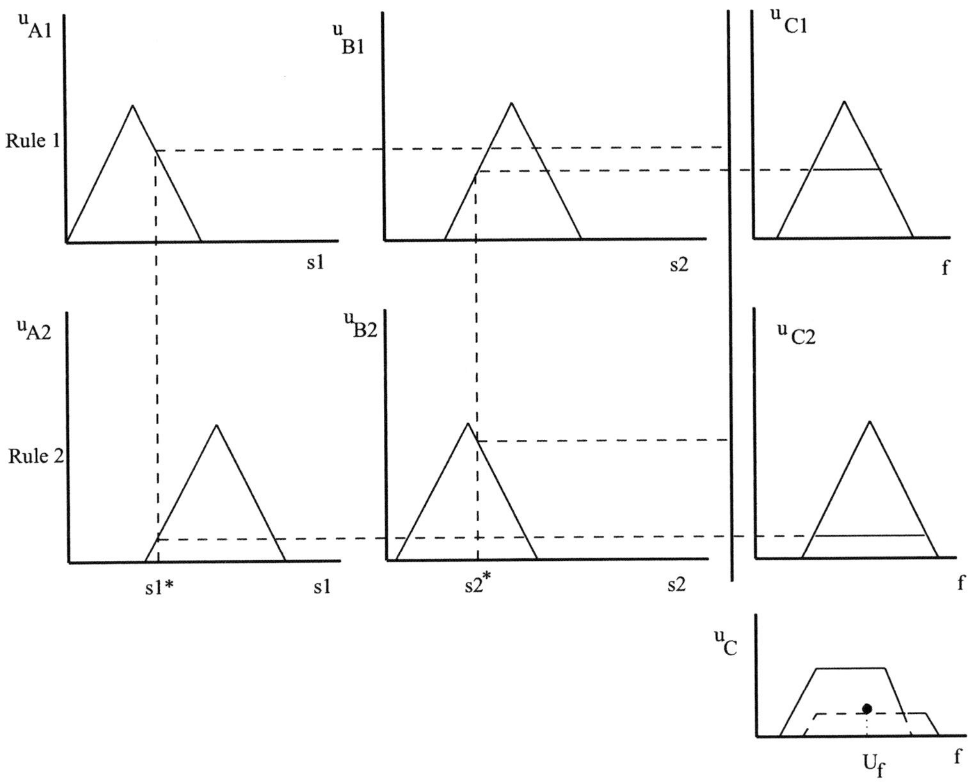

Figure 12: A schematic diagram showing the working principle of an FLC.

there are two fuzzy control rules as follows:

RULE 1: IF $s1$ is $A1$ and $s2$ is $B1$ THEN f is $C1$
RULE 2: IF $s1$ is $A2$ and $s2$ is $B2$ THEN f is $C2$.

If $s1^*$ and $s2^*$ are the inputs for fuzzy variables $s1$ and $s2$ and if u_{A1} and u_{B1} are the membership function values for A and B, respectively, then the grade of membership of $s1^*$ in $A1$ and the grade of membership of $s2^*$ in $B1$ are represented by $u_{A1}(s1^*)$ and $u_{B1}(s2^*)$, respectively for rule 1. Similarly, for rule 2, $u_{A2}(s1^*)$ and $u_{B2}(s2^*)$, are used to represent the membership function values. The firing strengths of the first and second rules are calculated as follows:

$$\alpha_1 = \min\left(u_{A1}(s1^*),\; u_{B1}(s2^*)\right), \quad (1)$$

$$\alpha_2 = \min\left(u_{A2}(s1^*),\; u_{B2}(s2^*)\right). \quad (2)$$

The membership function of the combined control action C is given by

$$u_C(f) = \max\left(u^*_{C1}(f),\; u^*_{C2}(f)\right). \quad (3)$$

The center of sums method may be employed for defuzzification and the method can be represented by

$$U_f = \frac{\sum_{j=1}^{p} A(\alpha_j) \times f_j}{\sum_{j=1}^{p} A(\alpha_j)}, \quad (4)$$

where U_f is the output of the controller, $A(\alpha_j)$ represents the firing area of the j-th rule, p is the total number of the fired rules and f_j represents the centroid of a membership function.

An FLC has the following advantages:

- It is a potential tool for dealing with imprecision and uncertainty.
- It does not require the extensive mathematical formulation of the problem.
- The rules are expressed in terms of – $IF \ldots THEN$ form. Thus, it becomes easier for the user to understand the control action.
- It might be computationally faster compared to other members of soft computing family, such as neural network, genetic algorithm, and others.

But, it suffers from the following drawbacks:

- The performance of an FLC depends mainly on its knowledge base (KB) and designing a proper KB of an FLC is not an easy task. The designer should have a thorough knowledge of the process to be controlled.
- The number of rules of an FLC depends on the number of variables and that of the grades/terms of each variable. Thus, the complexity of an FLC increases for handling the problems involving many variables.

Appendix B

B Neural Network

Artificial neural network (ANN) has been designed by copying the brain of a human-being artificially. Human brain contains a large number of computing elements (approximately 10^{11}) known as neurons and each neuron consists of a bush of thin fibers called dendrites, a long cylindrical fiber known as axon, synapse (a contact region of two neighboring neurons), and others.

Fig. 13 shows the schematic diagram of a biological neuron. A neuron collects information from its neighbors with the help of its dendrites. The collected information is then summed up in the cell body/ soma, before it is passed to the axon. This information is transferred to the next neuron through the synapse. These functions of a biological neuron are copied in an artificial neuron (refer to Fig. 14).

The function of a summing junction is similar to that of the combined dendrites and soma, and the activation/transfer function resembles the function of a synapse. It is important to mention that sometimes the output of the summing junction may become equal to

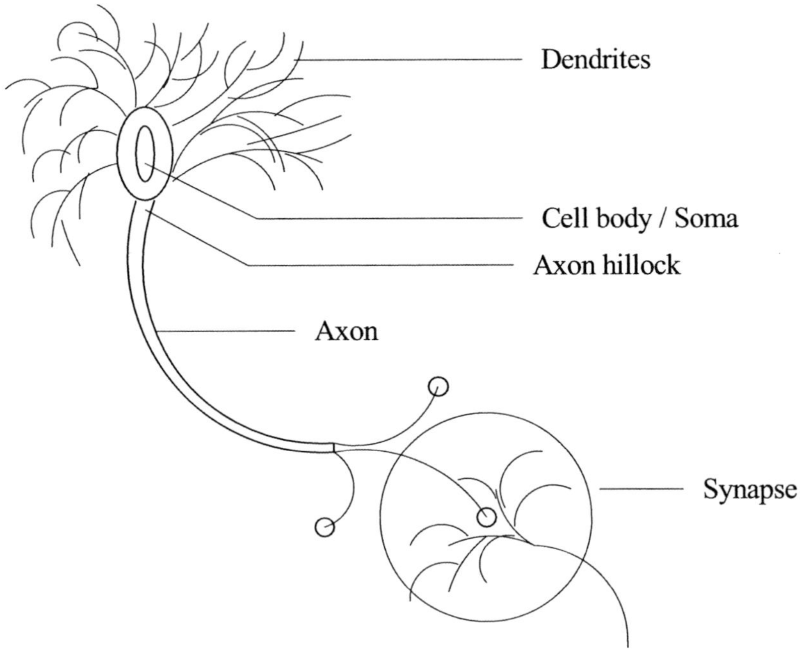

Figure 13: A schematic diagram showing a biological neuron.

zero, and to prevent that situation, a bias of fixed value is added to the summing junction. The output of k-th neuron (refer to Fig. 14) can be written as follows.

$$O_k = f(v_k) = f(\sum_{j=1}^{m} I_j w_{kj} + b_k) \tag{5}$$

In an ANN, the output of a neuron largely depends on its transfer/ activation function. Different types of transfer function are in use, such as hard-limit, linear, log-sigmoid, tan-sigmoid, and others. The neurons are connected by the weights and depending on the structure/topology of the networks, they are generally classified into several types, namely back-propagation NN, recurrent NN, self-organizing map, and others. The working principle of a back-propagation NN has been discussed briefly, in the next sub-section.

B.1 Back-Propagation NN

Fig. 15 shows the schematic diagram of a neural network, consisting of three layers, such as input, hidden and output layers. Inputs and output of the NN are denoted by $[I]$ and $[O]$, respectively. The $[V]$ and $[W]$ indicate the weights between the input & hidden layers and the hidden & output layers, respectively.

Initially, the weights – $[V]$ and $[W]$ are generated at random. Depending on the weights and transfer functions, the output of the network is determined, for a set of inputs. Each output is then compared to its corresponding target value and the error is calculated by using the equation given below.

$$E = \frac{1}{2}(T - O)^2, \tag{6}$$

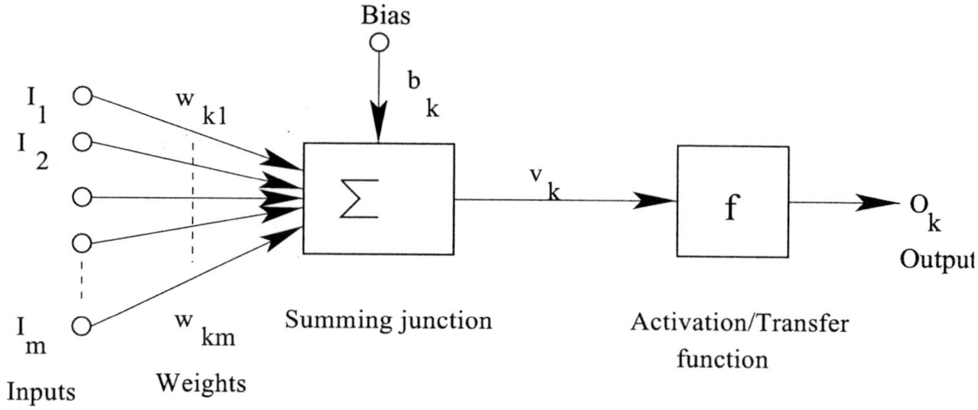

Figure 14: A schematic diagram showing an artificial neuron.

where T is the target value. This error E is then propagated in the backward direction, to modify the $[V]$ and $[W]$ values, so that the error becomes minimum, by using the following equations.

$$\Delta V = -\eta \frac{\partial E}{\partial V}, \tag{7}$$

$$\Delta W = -\eta \frac{\partial E}{\partial W}. \tag{8}$$

where η is the learning rate.

Training is generally provided to the NN, based on a pre-determined training set (i.e., in case of supervised learning), so that it can perform in an optimal sense. In supervised learning, training may be given to the network in two different ways, such as incremental training and batch training. Incremental training is also known as on-line training, in which the weights and biases of the network are updated each time one set of input (related to one training case) is passed to the network. In case of batch training, the weights and biases are updated after all sets of the input (belonging to all the training cases) have been presented to the network. It is also known as off-line training.

An NN has the following advantages:

- It can handle a control problem involving many variables.
- It may not require so much problem information, as that is needed for developing a fuzzy logic controller.

But, it has the following disadvantages:

- The solutions of a back-propagation NN may get stuck at local minima.
- Training of an NN is more involved computationally compared to tuning of a fuzzy logic controller.
- It works like a black box.

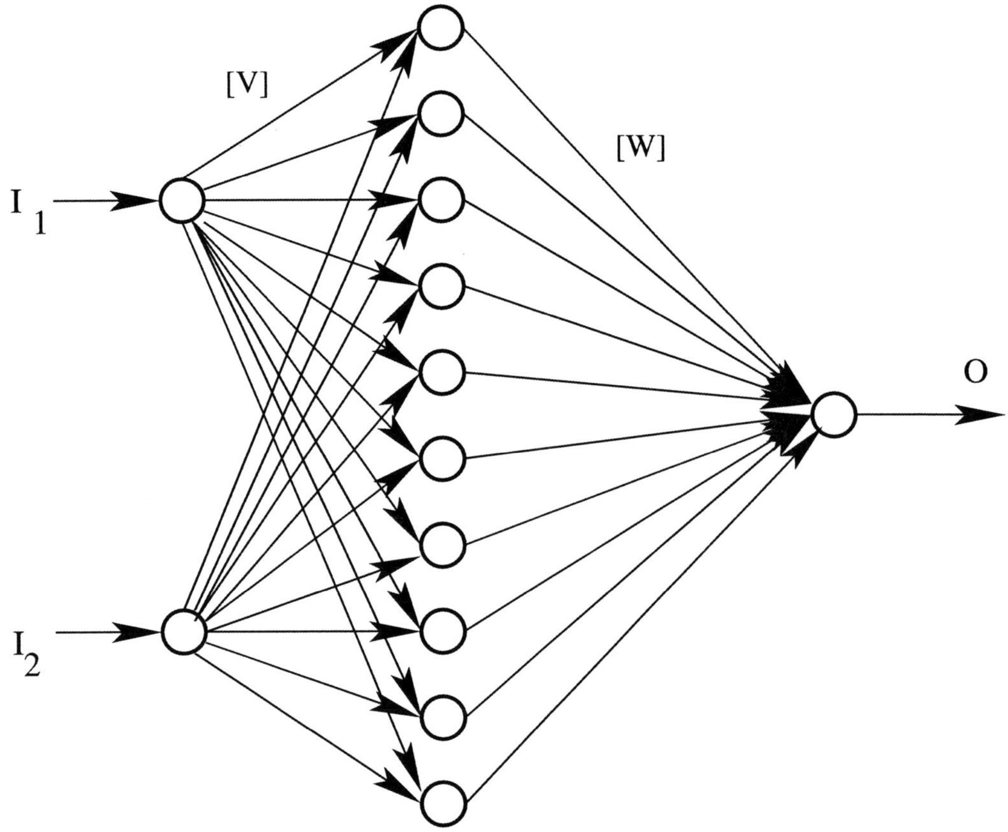

Input layer Hidden layer Output layer

I_1, I_2 : Inputs of the NN–controller

O : Output of the controller

[V], [W] : Weights of the NN–controller

Figure 15: A schematic diagram of a back-propagation neural network.

Appendix C

C Genetic Algorithm

Genetic algorithms (GAs) are population-based search and optimization algorithms, which mimic the principles of natural genetics and natural selection [61]. The GAs are very much different from classical search and optimization methods. There are several versions of GAs, such as binary-coded GA, real-coded GA, messy GA, micro-GA, and others, which are used to solve different kinds of search and optimization problems. The working principle of a binary-coded GA is explained below.

C.1 Binary-coded GA

In this type of GA, all problem variables are coded in finite-length binary strings. For example, three variables x_1, x_2, and x_3 can be represented by 4, 3, and 5 bit sub-strings as follows:

$$\underbrace{\texttt{0111}}_{x_1}\ \underbrace{\texttt{101}}_{x_2}\ \underbrace{\texttt{00110}}_{x_3}$$

The total string length (L) of a solution is then 12. The length of each sub-string is determined by the required accuracy in each variable. To retrieve the corresponding real values of the variables, the following linear mapping rule is used:

$$x_i = x_i^{(lower)} + \frac{x_i^{(upper)} - x_i^{(lower)}}{2^L - 1} \times d,$$

where d indicates decoded value of the sub-string and l is the length of the sub-string used to represent x_i. Once the values of the variables are known, objective function value can be determined. This objective function value is treated as the fitness value of the string.

The working principle of GA is shown in the form of a flowchart (refer to Fig. 16).

The operation of a GA begins with a population of random strings representing design variables. Thereafter, each string is evaluated to find the fitness value. The population is then operated by three main operators, namely reproduction, crossover, and mutation, to create a new population. The new population is further evaluated and tested for termination.

Reproduction operator selects good strings from the population using fitness information, to form a mating pool. There are several schemes for reproduction, namely proportionate selection, tournament selection, ranking selection, and others. In proportionate selection, the probability of a string for getting selected in a mating pool is proportional to the ratio of its fitness to the total fitness of the population. In a binary tournament selection, two strings (if the tournament size is assumed to be equal to two) are chosen at random from the population and the better string is selected and copied in the mating pool. This process continues by comparing two strings at a time till the mating pool has the same size as the original population size. Ranking selection is performed in two stages – in the first stage, the strings are sorted from the best to the worst according to their fitness values and the strings are copied according to a non-increasing assignment function in the second stage. Thus, this operator emphasizes the good strings of a population and make duplicate copies of them in a mating pool.

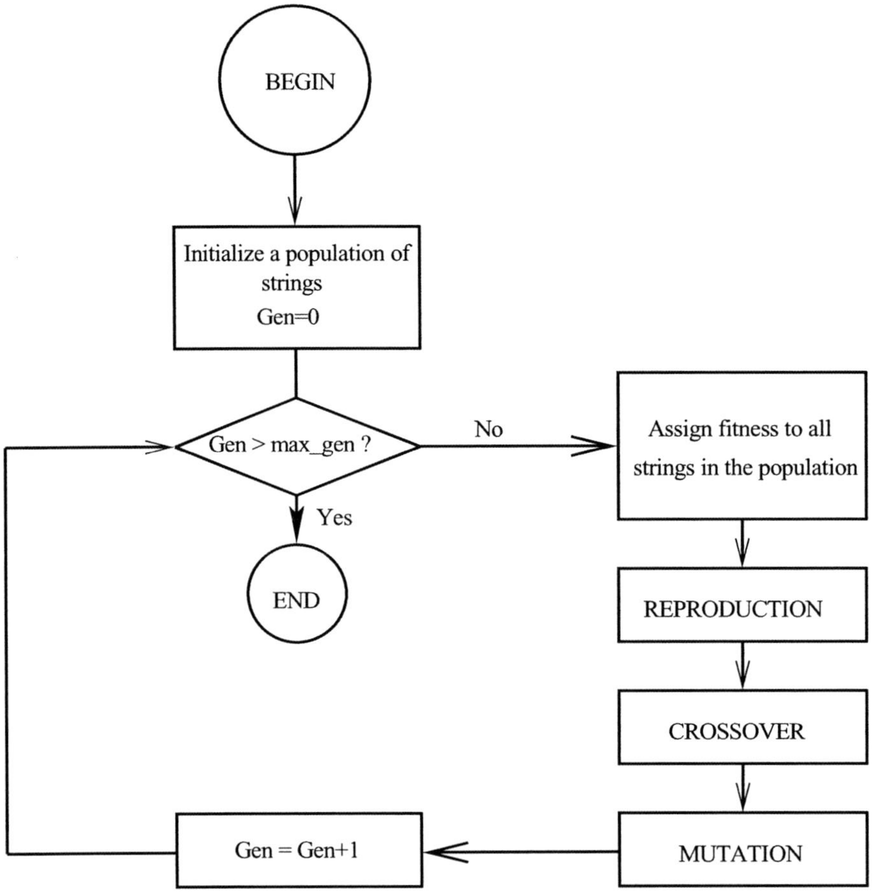

Figure 16: A schematic diagram showing the working principle of a GA.

In the crossover, new strings are created by exchanging information among strings of the mating pool. There exists a number of crossover schemes, such as single-point crossover, two-point crossover, multi-point crossover, uniform crossover, and others. In a single-point crossover, two strings are chosen from the mating pool, and also a crossing site is chosen at random along the string. Thereafter, all bits on the right side of the crossing site are exchanged between both the strings. In two-point crossover, two crossover sites are selected at random along the length of the parent strings and the bits lying between the two crossover sites are swapped between themselves. In case of multi-point crossover, a number of crossover sites (generally equal to the number of variables represented by the string) are chosen at random and the bits are exchanged at those sites between the two parents following the principle of single-point crossover. Uniform crossover is generally performed with a probability of 0.5. A coin is flipped with a probability of 0.5 for appearing head at each bit-position of the strings. If the head appears, the bits at that particular position are swapped, otherwise those are kept unaltered. This process continues, till all the bit-positions along the strings are checked for a possible crossover to occur. This operator allows partial information to be exchanged between two good strings found using the reproduction operator.

Crossover operator is mainly responsible for the search of new strings. In order to reduce the chance of destructing already-found good strings, crossover is usually performed with a probability p_c almost near to one.

Mutation operator changes 1 to 0 and vice versa with a small probability, p_m. Mutation is used for achieving a local change around the current solution. In short, reproduction operator selects good strings and crossover operator recombines two good strings to hopefully create better strings. The mutation operator alters a string locally to create a new string.

The above operation of a GA is very similar to the evolutionary principle. If good strings are created by crossover and mutation operators, reproduction emphasizes them and they have more chance of getting mated with other good solutions. On the other hand, if crossover and mutation creates bad strings, reproduction ruthlessly eliminates them from further processing. The string copying and sub-string exchange operations in GAs may seem at first to be random operators, but a careful study will prevail that although random numbers are used extensively in a GA, the search is not random. Instead, the randomness in the search operators provide the necessary stochasticities for a GA not to get stuck at sub-optimal solutions.

The advantages of a GA have been listed below.

- It does not require gradient information of a function to be optimized.
- The GA starts with a population of solutions. Thus, the chance of its solutions for getting trapped into a local minimum is less.
- A GA can solve a variety of optimization problems. It is versatile and more robust compared to conventional optimization tools.

But, it suffers from the following disadvantages:

- It is computationally expensive and has a slow convergence rate.
- The performance of a GA depends on its parameters and the user should have a thorough knowledge of the grammar of a GA, to choose an appropriate set of parameters.

References

[1] Mackworth A. On seeing robots, *Computer Vision: Systems, Theory and Applications*, World Scientific Press, Singapore, 1, 1993.

[2] Asada M., Kitano H., Noda I., Veloso M. RoboCup: today and tomorrow – what we have learned, *Artificial Intelligence*, **110**, 193–214 (1999).

[3] Shibata T. and Fukuda T. Coordination in evolutionary multi-agent robotic system using fuzzy and genetic algorithm, *Control Engineering Practice*, **2**,**1**, 103–111 (1994).

[4] Jaeger H. and Christaller T. Dual dynamics: designing behavior systems for autonomous robots, *Artificial Life and Robotics*, **2**, 108–112 (1998).

[5] Zadeh L.A. The role of soft computing and fuzzy logic in conception, design and development of intelligent systems, *Proc. of Intl. Workshop on Soft Computing in Industry*, Muroran, Japan, 136–137 (1996).

[6] Zadeh L.A. Foreword, *Proc. of second International Conference on Fuzzy Logic and Neural Networks*, Iizuka, Japan, XIII–XIV (1992).

[7] Merriam-Webster *Webster's Ninth New Collegiate Dictionary*, Massachusetts, USA: Springfield, 1985.

[8] Latombe J.C. *Robot Motion Planning*, Boston, MA: Kluwer Academic Publishers, 1991.

[9] Nilsson N.J. A mobile automation: an application of artificial intelligence techniques, *Proc. of the First International Joint Conference on Artificial Intelligence*, Washington D.C., USA, 509–520 (1969).

[10] O'Dunlaing C., Sharir M., Yap C. Generalized Voronoi diagram for a ladder I: Topological analysis, *Comm. Pure Appl. Math.*, **39**, 423–483 (1986).

[11] Brooks R.A. Solving the find-path problem by good representation of free space, *IEEE Trans. on Systems, Man and Cybernetics*, **13, 3** 190–197 (1983).

[12] Lozano-Pérez T. Spatial planning: a configuration space approach, *IEEE Trans. on Computers*, **C-32, 2**, 108-120 (1983).

[13] Liu Y.H. and Arimoto S. Path planning using a tangent graph for mobile robots among polynomial and curved obstacles, *International Journal of Robotics Research*, **11, 4** 376–382 (1992).

[14] Fujimura K. and Samet H. Accessibility: a new approach to path planning among moving obstacles, *Proc. of IEEE Conf. on Computer Vision and Pattern Recognition*, Ann Arbor, USA, 803–807 (1988).

[15] Khatib O. Real-time obstacle avoidance for manipulators and mobile robots, *The International Journal of Robotics Research*, **5, 1** 90–98 (1986).

[16] Kant K. and Zucker S.W. Towards efficient planning: The path velocity decomposition, *The International Journal of Robotics Research*, **5** 72–89 (1986).

[17] Sharma R. A probabilistic framework for dynamic motion planning in partially-known environments, *Proc. of IEEE Int. Conf. on Robotics and Automation*, Nice, France, 2459–2464 (1992).

[18] Zhu Q. Hidden Markov model for dynamic obstacle avoidance of mobile robot navigation, *IEEE Trans. on Robotics and Automation*, **7, 3** 390–397 (1991).

[19] Nam Y.S., Lee B.H., Kim M.S. View-time based moving obstacle avoidance using stochastic prediction of obstacle motion, *Proc. of IEEE Int. Conf. on Robotics and Automation*, Minneapolis, Minnesota, USA, 1081–1086 (1996).

[20] Fiorini P., Shiller Z. Motion planning in dynamic environments using the relative velocity paradigm, *Proc. of IEEE Conf. on Robotics and Automation*, 560–565 (1993).

[21] Brooks R.A. A robust layered control system for a mobile robot, *IEEE Jl. of Robotics and Automation*, **RA-2** 14–23 (1986).

[22] Reif J. and Sharir M. Motion planning in the presence of moving obstacles, *Proc. of IEEE Symp. on Foundations of Computer Science*, 144–154 (1985).

[23] Canny J. and Reif J. New lower bound techniques for robot motion planning problems, *Proc. of IEEE Symp. on Foundations of Computer Science*, **49**–60 (1987).

[24] Pratihar D.K. Algorithmic and soft computing approaches to robot motion planning, *Machine Intelligence and Robotic Control*, **5, 1** 1–16 (2003).

[25] Zadeh L.A. Fuzzy sets, *Information and Control*, **8** 338–353 (1965).

[26] Beaufrere B. and Zeghloul S. A mobile robot navigation method using a fuzzy logic approach, *Robotica*, **13** 437–448 (1995).

[27] Takeuchi T., Nagai Y., Enomoto Y. Fuzzy control of a mobile robot for obstacle avoidance, *Information Sciences*, **45, 2** 231–248 (1988).

[28] Martinez A., Tunstel E., Jamshidi M. Fuzzy logic based collision avoidance for a mobile robot, *Robotica*, **12** 521–527 (1994).

[29] Pin F.G. and Watanabe Y. Navigation of mobile robots using a fuzzy behaviorist approach and custom- designed fuzzy inferencing boards, *Robotica*, **12** 491–503 (1994).

[30] Maeda M., Shimakawa M., Murakami S. Predictive fuzzy control of an autonomous mobile robot with forecast learning function, *Fuzzy Sets and Systems*, **72, 1** 51–60 (1995).

[31] Kim J.H. and Kong S.G. Fuzzy systems for group intelligence of autonomous mobile robots, *Proc. of the IEEE International Fuzzy Systems Conference*, Seoul, Korea, I-100–I-104 (1999).

[32] Marapane S.B., Trivedi M.M., Lassiter N., Holder M.B. Motion control of cooperative robotic teams through visual observation and fuzzy logic control, *Proc. of IEEE International Conference on Robotics and Automation*, Minneapolis, Minnesota, USA, 1738–1743 (1996).

[33] McCulloch W.S. and Pitts W. A logical calculus of the ideas immanent in nervous activity, *Bulletin of Mathematical Biophysics*, **5**, 115–133 (1943).

[34] Lee J. and Bein Z. Collision-free trajectory control for multiple robots based on neural optimization network, *Robotica*, **8, 3** 185–194 (1990).

[35] Botelho S.S.C., Valle Simoes E., Uebel L.F., Barone D. High speed neural control for robot navigation, *Proc. of IEEE International Conference on Systems, Man and Cybernetics*, Beijing, China, **3** 1956–1959 (1996).

[36] Holland J.H. *Adaptation in Natural and Artificial Systems*, Ann Arbor: The University of Michigan Press, 1975.

[37] Mitchell J.S.B. An algorithmic approach to some problems in terrain navigation, *Artificial Intelligence*, **37** 171–201 (1988).

[38] Zhao M., Ansari N., Hou E.S.H. Mobile manipulator path planning by a genetic algorithm, *Journal of Robotic Systems*, **11, 3** (1994).

[39] Pratihar D.K. *Path and Gait Generation of Legged Robots Using a GA-Fuzzy Approach*, Ph.D. thesis, IIT Kanpur, India (2000).

[40] Pratihar D.K. and Bibel W. Time-optimal, collision-free path planning for multiple cooperating robots using a genetic-fuzzy system, *Machine Intelligence and Robotic Control*, **5, 2** 45–58 (2003).

[41] Li W., Ma C., Wahl F.M. A neuro-fuzzy system architecture for behavior-based control of a mobile robot in unknown environments, *Fuzzy Sets and Systems* **87, 2** 133–140 (1997).

[42] Ng K.C. and Trivedi M.M. A neuro-fuzzy controller for mobile robot navigation and multi-robot convoying, *IEEE Trans. on Systems, Man and Cybernetics, Part B* **28, 6** 829–840 (1998)

[43] Topalov A.V., Tsankova D.D., Petrov M.G., Proychev T.P. Intelligent sensor-based navigation and control of mobile robot in a partially known environment, *Proc. of 3rd IFAC Symposium*, Madrid, Spain, 267–272 (1998).

[44] Acosta L., Marichal G.N., Moreno L., Méndez J.A., Rodrigo J.J. Obstacle avoidance using the human operator experience for a mobile robot, *Journal of Intelligent and Robotic Systems: Theory and Applications*, **27, 4** 305–319 (2000).

[45] Pratihar D.K. Evolutionary robotics – a review, *Sadhana*, **28, 6** 999–1009 (2003)

[46] Floreano D. and Mondada F. Automatic creation of an autonomous agent: Genetic evolution of a neural network driven robot, *Proc. of the 3rd Conference on Simulation of Adaptive Behavior*, Cambridge, MA:The MIT Press (1994).

[47] Nolfi S., Miglino O., Parisi D. Phenotypic plasticity in evolving neural networks: evolving the control system for an autonomous agent, *Technical Report PCIA-94-04*, Institute of Psychology, CNR, Rome, Italy (1994).

[48] Harvey I., Husbands P., Cliff D., Thompson A., Jakobi N. Evolutionary robotics: The Sussex Approach, *Robotics and Autonomous Systems*, **20** 205–224 (1997).

[49] Krishnakumar K. Micro-genetic algorithms for stationary and nonstationary function optimization, *Proc. of SPIE International Conference on Adaptive Systems*, Philadelphia, USA, 289–296 (1989).

[50] Alami R., Fobert F., Ingrand F., Suzuki S. Multi-robot cooperation through incremental plan-merging, *Proc. of IEEE International Conference on Robotics and Automation*, **2573**–2579 (1995).

[51] Svestka P., Overmars M.H. Coordinated motion planning for multiple car-like robots using probabilistic roadmaps, *Proc. of IEEE International Conference on Robotics and Automation*, **1631**–1636 (1995).

[52] Barraquand J., Langlois B., Latombe J.C. Numerical potential field techniques for robot path planning, *IEEE Trans. on Systems, Man, and Cybernetics*, **22, 2**, 224–241 (1992).

[53] Vail D., Veloso M. Dynamic multi-robot coordination, in Schultz A., Parker L., Schneider F. (Eds) *Multi-robot Systems: From Swarms to Intelligent Automata*, Vol. II, Kluwer Academic Publishers, 87–100 (2003).

[54] Fukuda T. Intelligent, interactive robot with emotion, *HUMBOLDT Kosmos*, **83**, 31 (2004)

[55] Pratihar D.K., Deb K., Ghosh A. A genetic-fuzzy approach for mobile robot navigation among moving obstacles, *International Journal of Approximate Reasoning*, **20** 145–172 (1999).

[56] Pratihar D.K., Deb K., Ghosh A. Design of genetic-fuzzy systems for planning crab gaits of a six-legged robot, *Journal of Computing and Information Technology, Special issue on Evolutionary Computation*, **7, 1** 93–101 (1999).

[57] Pratihar D.K., Deb K., Ghosh A. Optimal turning gait of a six-legged robot using GA-fuzzy approach, *Artificial Intelligence for Engineering Design, Analysis and Manufacturing - AIEDAM*, **14** 207–219 (2000).

[58] Pratihar D.K., Deb K., Ghosh A. Optimal path and gait generations simultaneously of a six-legged robot using a GA-fuzzy approach, *Robotics and Autonomous Systems*, **41, 1** 1–20 (2002).

[59] Jha R.K., Singh B. and Pratihar D.K. On-line stable gait generation of a two-legged robot using a genetic-fuzzy system, *Robotics and Autonomous Systems*, **53** 15–35 (2005).

[60] Mamdani E.H. and Assilian S. An experiment in linguistic synthesis with a fuzzy logic controller, *International Journal of Man-Machine Studies*, **7** 1–13 (1975).

[61] Goldberg D.E. *Genetic Algorithms in Search, Optimization, and Machine Learning*, Reading, Mass., USA: Addison-Wesley, 1989.

INDEX

D

E

F

G

H

I

J

K

L

M

N

O

P

Q

R

S

T